T0296190

HUMORAL AGENTS IN NERVOUS ACTIVITY

HUMORAL AGENTS IN NERVOUS ACTIVITY

WITH SPECIAL REFERENCE TO
CHROMATOPHORES

BY

G. H. PARKER

Harvard University

CAMBRIDGE
AT THE UNIVERSITY PRESS
1932

CAMBRIDGE
UNIVERSITY PRESS

University Printing House, Cambridge CB2 8BS, United Kingdom

Cambridge University Press is part of the University of Cambridge.

It furthers the University's mission by disseminating knowledge in the pursuit of education, learning and research at the highest international levels of excellence.

www.cambridge.org
Information on this title: www.cambridge.org/9781107502291

First published 1932
First paperback edition 2015

A catalogue record for this publication is available from the British Library

ISBN 978-1-107-50229-1 Paperback

To
James Gray

CONTENTS

CONTENTS

PREFACE

THE idea prevalent among the ancients that personality and the control of bodily responses were aspects of man's nature that permeate his whole frame gave way gradually with the increase of knowledge to the view that these functions belong exclusively to his nervous system, a doctrine that was at its height in the days of Cuvier about a century ago. Some decades after this period Claude Bernard called attention to the great importance in this respect of the internal fluids of the body with their immensely complex array of solutes. These collectively constituted in his opinion an internal environment for the organism quite as important as the external one and of first significance in understanding any creature's economy. It was not, however, till 1902 that Bayliss and Starling initiated an analytic attack on this field by the discovery of secretin and by introducing the conception of the hormone. Thus was brought about a new method of approach to this aspect of organic activity, and bodily control by hormones was set up in strong contrast with that by nerves. In a rather remote way the new doctrine revived the ancient one of the cardinal humours.

In the first edition of the *Principles of General Physiology* the distinguished author, Professor Bayliss, raises the question as to the relative ages of the nervous and the humoral control; he asks, in other words, which of these two contrasted systems of co-ordination appeared first in the course of animal evolution. Professor Bayliss was inclined on the whole to favour hormones as the more primitive agents. That the relations of these two systems are of a somewhat different nature from that suspected by him and far more intimate than he ever supposed them to be is the conclusion to be drawn from the present study, a conclusion, however, that is rather a working hypothesis than a final answer.

The content of this volume is the substance, somewhat expanded and brought up to date, of a lecture that the writer was kindly invited to give in May, 1930, to the members of the Zoological Laboratory of the University of Cambridge. He wishes to express to the Laboratory Staff his appreciation of the honour done him in the invitation. He also wishes to thank the Syndics of the University Press of Cambridge for their generous suggestion that the manuscript be submitted to them for publication. Acknowledgments are due to Miss Mary F. Wolfe, Miss Alice Starkey, and Mrs G. H. Parker for their care in the preparation of the text, and Dr F. M. Carpenter and Mr E. A. Schmitz for their kind attention to the illustrations.

G. H. PARKER

Harvard University
June 1931

CHAPTER I

INTRODUCTION

I N the more complex animals the nervous system with its appended parts falls into three sharply separated sets of organs: receptors as represented by what are commonly called sense organs, adjustors or central nervous organs, and effectors or organs of response. The receptors include such parts as the eye, the ear, organs of smell and of taste and the like, all of which in consequence of the pronounced and characteristic sensations with which they are associated have been rightly catalogued as sense organs. But under receptors are also included other organs such as those of equilibrium in the ear and the proprioceptors of the muscles, most of which, under normal circumstances, give rise to no clear and definite sensations and consequently are scarcely to be regarded in strict terminology as organs of sense. Such receptors are primarily concerned with the excitation of responses in the animals that possess them rather than with the production of sensations. At least this is what we may infer from our own experience. Notwithstanding the unobtrusive character of receptors of this kind, they are often of first importance for the welfare of the creature in which they occur. They are associated more distinctly with what may be called the organic side of animal life, than with what has to do with mentality. They represent the initial phase in the evolution of receptors in general, for in many of the simpler animals they are the only kind present. This opinion is justified by the fact that in many instances in the coelenterates the receptors are directly connected with muscles and are in no way associated with central organs. Such receptors are means of exciting action and nothing else. In some of these simpler animals, as for instance certain jelly-fishes, receptors highly enough specialized to have been called eyes are thus related to the musculature of their possessors. Animals of this kind are not to be described as having a sense of sight; they may be said merely to respond to light. Notwithstanding this limitation, such receptors enable the animals that possess them to meet successfully the changes in their environment. These animals respond much as our organs of digestion or of circulation do in meeting the exigencies placed upon them. When we change from walking on the level to climbing a hill, the heart and other elements in our circulatory system readjust themselves in what is essentially an automatic manner, quite independent of our volition. Nevertheless, this readjustment involves receptor activities as an essential part of its accomplishment. The nervous life of a simple animal whose receptors are exclusively of this type may be described as organic, in contrast with the mixed state seen in the more complex creatures such as ourselves. Here, in addition to almost purely organic receptors, are others, true sense organs, which, though they may excite responses of one kind or another, also call forth in us those sensational elements that serve as the basis of our intellectual life.

The adjustors or central nervous organs are represented in the vertebrates by the brain, spinal cord and the autonomic nervous system, and in the invertebrates by various ganglionic enlargements usually designated as brain and ventral chain. All these central organs are outgrowths from the nerve-net of the simpler invertebrates. In this net, as seen in most coelenterates, the nerve cells or protoneurones are so connected that nervous impulses pass over them in any direction. Such nets are structurally and functionally unpolarized. In the central nervous organs of the more complex animals, on the other hand, conduction of impulses occurs in only specified directions. Here the protoneurones of the nerve-net have become differentiated into neurones, highly specialized nerve cells whose interrelation is through synapses by which impulses are allowed to pass in one direction only. Hence the elements in this type of organ are polarized and transmission is directed toward specific ends. In such a differentiated central organ conduction is not only restricted in direction, but it is also so circumscribed that under normal conditions particular receptor fields become associated with particular sets of effectors. In this way the central organ comes to extend a limiting influence over the final response.

The flow of nervous impulses, purely momentary and immediate in its outcome in the simpler creatures, acquires in the more specialized forms the capacity to leave in the central organs an impress of its passage and in this way to influence more or less all subsequent nervous activity. An animal such as a sea-anemone, with a relatively simpler organization, will, if plied vigorously, continue to take in food till it is gorged or, if it is provided with an artificial outlet, it will pass through its body without cessation food to the extent of many times its own volume. Complex animals, on the other hand, ordinarily cease taking food before they have reached a physical limit. In other words, with them the early steps in food taking influence their nervous organization in such a way that the later steps in this process are distinctly modified; the process, in fact, may be thereby brought to a standstill. In this way the central organs of the more complex animals retain for longer or shorter periods evidence of change in the creature's state and thus become repositories of its experience. Such operations lay the foundations for what we recognize in ourselves as the beginnings of the intellectual life.

Effectors are the parts by which animals respond to the changes in the world about them. Of the various types included under this term muscles are the most commonly known. But in addition to muscles animals possess such effectors as urticating organs, chromatophores, luminous organs, glands, cilia and electric organs, to name only those that may be designated as elementary. Urticating organs are represented by the nettling cells of the jelly-fishes and other members of the Coelenterata, all of which are characterized by the possession of these microscopic irritants. Chromatophores are the integumentary cells by which a number of animals such as certain crustaceans, fishes, amphibians and reptiles, may make temporary changes in the colours of their skins. Luminous organs are well known in numerous marine animals from the protozoans to the fishes; they produce the so-called phosphorescence of the sea. On land they are well developed in the fire-flies. Glands are the organs of secretion throughout the whole range of the animal kingdom. Cilia, using that term in its broadest sense, are microscopic, protoplasmic lashes which by their active beating either move small aquatic animals through a watery medium or generate currents in the fluids that bathe the surfaces or occupy the cavities of the larger creatures. Cilia are to be found in all groups of the animal kingdom excepting possibly the thread-worms and the arthropods. Muscle is the universal motor element of all animals, and electric organs, confined to a few fishes, are modified muscles in which motility has been superseded by the capacity to generate electricity. These elementary effectors singly or in combination are the means whereby animals react to the environmental changes about them. They are the organs ordinarily set into activity by the flood of nervous impulses that originate in the receptors, and reach them through the paths of the central organs.

Elementary effectors are brought into action in one or both of two ways. Either they are excited to respond by the impinging of external changes directly upon them or they are brought into action through nerve impulses that, in a sense, represent such changes indirectly. Elementary urticators are apparently always activated through direct stimulation, for, notwithstanding statements to the contrary, no conclusive evidence has ever been brought forward to show that they are under the control of nerves. All other types of effectors, excepting electric organs, are known to be excitable in one instance or another by both direct stimulation and nervous impulse. Electric organs are apparently never normally stimulated except through nerves. This results from the fact that in the fishes possessing them the particular muscles from which they were derived were, before their transformation into electric organs, under nerve control. Hence the electric organ from its very beginning was supplied with nerves.

No single animal possesses the full range of elementary effectors. In fact, most animals, as the instances given in Table I indicate, possess only three or four sets

Table I

Distribution of elementary effectors among six common animals in range from simple to complex

	Sea-anemone	Earth-worm	Fire-fly	Torpedo	Frog	Man
Urticating organs	×	—	—	—	—	—
Glands	×	×	×	×	×	×
Luminous organs	—	—	×	—	—	—
Chromatophores	—	—	—	×	×	—
Cilia	×	×	—	×	×	×
Muscles	×	×	×	×	×	×
Electric organs	—	—	—	×	—	—

of such organs. Of the animals recorded in this table, the torpedo has the highest number, five, and the earthworm, the fire-fly, and man the lowest, three. It is interesting to note that the earthworm and man are not only limited to three sets of effectors, but to the same three, glands, cilia, and muscles. Notwithstanding the enormous complexity that subsists between man and his environment, these three sets of organs appear to be all that are necessary to meet such conditions successfully. In fact, when we recall the chief elements of our effector relations with the environment we are forced to admit that we meet this situation chiefly through only one set of these organs, namely, muscles. Our cilia are mainly concerned with the removal of refuse from the cavities of our bodies in that these effectors generate currents that pass from the depths of these cavities out toward the exterior and thus relieve them of any accumulated waste. In this way the sinuses of the head, the lungs, certain digestive spaces, and the urogenital tracts are kept free from noxious accumulations. Most glands have to do with our internal economy and are not specifically concerned with the outer world. But tears may express emotion and expectoration plays a part in certain phases of our life. Even the scent glands of our skin must be reckoned with, but in none of these instances do the effectors concerned approximate in importance to our muscles. By means of these organs we accomplish something prodigiously more than with any of our other effectors. By muscles we produce all the varied movements of our bodies, locomotion and what man can do with his hands from digging a ditch to playing a violin or modelling a statue. Muscles, too, give us facial expression and all those thousands of slight and almost imperceptible motions that express inward character. Most of all, muscles give us speech from the inarticulate cry to the most finished form of language. Muscle is the artist's tool and serves not only as his means of portraying to us his ideal worlds, but enables him on the stage to arouse in us a wealth of self-imposed emotions, none the less telling in consequence of their artificiality. These and a host of other actions we owe to our muscles. They are our supreme effectors. How great the contrast between what muscle means in us and in an earthworm!

The evolutionary history of the several classes of effectors is not without interest. Aside from the elementary urticating organs which are strictly limited to the coelenterates, and the electric organs which are peculiar to fishes, all other elementary effectors except chromatophores seem to have appeared very early in animal evolution and quite independent of nervous control. These independent effectors, luminous organs, glands, cilia, and muscles give indubitable evidence of early and separate origin. So far as their importance in the subsequent history of animals is concerned, muscle, as we have seen, far outran the others. Around luminous organs, glands, and cilia, no particular organic development ever took place. Whereas around muscle nervous tissue took its origin. Muscle in sponges is apparently always under direct stimulation, but in coelenterates it is commonly brought into action through groups of receptor cells which transmit directly to it impulses arising from environmental changes. Here no central organ intervenes. In the more complex animals central nervous organs occupy an intermediate place between the receptor and the muscle and these central organs arose as a later growth after effectors and receptors had been differentiated. Thus, adjustors were the third and last element to be added to this expanding complex, and served as a soil from which sprang a most remarkable set of activities, the central nervous functions. It is the purpose of the following pages to consider the mutual relations of these several constituents of the nervous system both in their fully developed condition and in their evolution, and to discuss in particular the importance of secretion as a controlling factor in the various operations, especially those shown by chromatophores (Parker, 1909, 1910, 1919, 1923; Fortuyn, 1920; Hanström, 1928; Kappers, 1929).

CHAPTER II

VERTEBRATE CHROMATOPHORES

CHROMATOPHORES, unlike most other effectors, are found in only the more complex animals. They are limited to such forms as have well-developed central nervous organs and highly differentiated eyes. They are to be met with in the cephalopods, in certain crustaceans, and in many of the cold-blooded vertebrates. In the cephalopods they are so complicated as to justify their classification under the head of simple

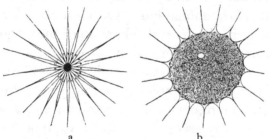

a   b

Fig. 1. Chromatophores of the squid *Loligo*: *a*, contracted; *b*, expanded. Bozler, *Zeit. vergl. Physiol.* 1928, **7**, 381, fig. 1.

organs rather than under that of cells, or even of cell aggregates. In these animals each chromatophore consists of two parts, a small central cell, spherical in form, containing within its elastic envelope an appropriately coloured fluid, and a circlet of numerous radiating smooth-muscle fibres provided with nerves. The ends of the muscle fibres are attached on the one hand to the envelope of the central cell and on the other to some point in the adjacent tissue (Fig. 1). When these fibres contract, they draw out the cell into a thin, flat disk whose exposed surface is very many times that which the cell possessed in its spherical form. With this expansion the coloured contents of the sphere come to cover a relatively large area, and thus the chromatophore, in conjunction with many others of a like kind, contributes to the momentary colour of the animal's skin. Several such colour systems of different tints, and more or less independent in action, may exist side by side in the integument of the same animal. When the muscle fibres attached to the central cell relax, this cell, in consequence of the elasticity of its envelope, contracts from its flattened shape to that of a sphere again. It now assumes the appearance of a small coloured dot imbedded in the skin of the animal and it is so minute as to have little or no effect in the general coloration of the integument. The changes in such chromatophores as these, then, are the results of the action of smooth-muscle fibres on the central cells and the physiology of this type of chromatophore is, therefore, chiefly that of a smooth-muscle system.

In the other types of chromatophores, those found in the crustaceans and in the cold-blooded vertebrates, the mechanism of change is quite different. Here the chromatophores are either single cells or small groups of cells whose protoplasm contains coloured material. This material is either moved by protoplasmic streaming from the centre of the cell outward and back again, thus giving the appearance of expansion and contraction, or it is alternately generated and destroyed locally. By one or both of these methods the colour of the skin of the given animal is changed. The method of migration of coloured particles is more commonly exemplified in the fishes, amphibians, and reptiles, that of alternate generation and destruction in the crustaceans. It is from these simpler types of chromatophores rather than from those in the cephalopods that results concerning nervous secretions have been obtained.

The colour changes in fishes were known to the ancients and were referred to by both Aristotle and Pliny. These changes were made the object of experimental study about a century ago by Stark (1830), who

Fig. 2. Turbots in which the chromatophores have been rendered inoperative by the cutting of particular nerves. Pouchet, *Jour. Anat. Physiol.* 1876, **12**, pl. 4.

pointed out that by their means a fish could accommodate itself to the background on which it lived. Pouchet (1872*a*, 1872*b*, 1876) undertook a physiological investigation of these changes and showed that when certain nerves were cut, particularly those containing sympathetic fibres, the chromatophores of the regions supplied by these nerves took on an intermediate tint and remained, with but little change, in this state (Fig. 2).

These results were abundantly confirmed and greatly extended by von Frisch (1910, 1911*a*, 1911*b*, 1912), who not only worked out the peripheral sympathetic innervation, but determined the location in the brain and in the spinal cord of the controlling centres. Wyman (1924) devised a method for cutting a group of nerves supplying the caudal fin of a fish without, however, disturbing the blood supply to that region (Fig. 3). In *Fundulus* after this operation the portion of the tail thus denervated will maintain a uniform dark grey which commonly stands

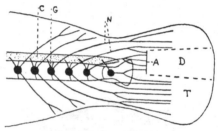

Fig. 3. Wyman's method of denervating part of the tail of a fish (*Fundulus*) without excluding the circulation of the blood. The transverse incision indicated at *A* establishes two areas in the tail, one denervated, *D*, and the other normal, *T*. *C*, spinal cord; *G*, spinal ganglion; *N*, spinal nerves. Wyman, *Jour. Exp. Zool.* 1924, **39**, 83, fig. 3.

out in contrast with the tint of the rest of the body. The general surface of such a fish will continue to exhibit the usual changes in response to environmental differences of illumination and thus will be at one time lighter, at another darker, than the denervated area. All such changes are dependent upon the skin melanophores, which in the dark condition are fully expanded, in the denervated state partly expanded (stellate stage), and in the light condition fully contracted (punctate stage) (Fig. 4). In all fishes thus far tested such changes have been shown to depend upon the eye. When fishes are blinded either by covering the eyes or excising them, responses to the tint of the environment cease. One eye is, however, sufficient to enable a fish to carry out the whole range of these responses. In a histological study of the melanophores in fishes Ballowitz (1893) long ago demonstrated an abundant supply of nerves

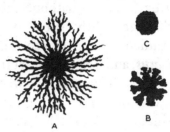

Fig. 4. Melanophores of *Fundulus* in three stages: *A*, fully expanded; *B*, stellate; *C*, fully contracted, punctate.

to these cells (Fig. 5). Thus histological studies as well as physiological experiments have demonstrated beyond a doubt that at least the melanophores of fishes are under direct nerve control, a conclusion conceded by almost every modern investigator of the subject.

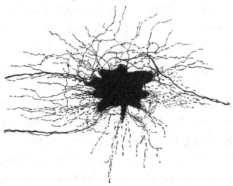

Fig. 5. Innervation of a chromatophore from the perch. Ballowitz, *Zeit. wiss. Zool.* 1893, **56**, pl. 38, fig. 21.

The number of kinds of chromatophores possessed by fishes varies more or less with the species. Some forms, as for instance the catfish *Ameiurus*, have apparently only one kind of such cells, the melanophore; consequently the changes in tint of this fish extend only from light grey to black, without exhibiting at any stage a phase that can rightly be called colour. In a number of fishes studied by von Frisch (1911 *a*, 1912), *Phoxinus*, *Crenilabrus*, and *Trigla*, at least three kinds of chromatophores, black, yellow, and red, were identified. By experiments involving nerve cutting von Frisch showed that all three kinds of colour cells were under the direct control of nerves. If the centre for colour control in the spinal cord of a freshly prepared *Crenilabrus* is stimulated, all three sets of chromatophores contract at once, thus giving evidence of a uniform system of innervation. In *Fundulus* three types of chromatophores have also been distinguished: the usual black melanophores, yellow xanthophores, and iridescent bluish green bodies or iridophores. Whether the iridophores can change or not and, if they do change, what effect such change may have on the general coloration of the fish is still to be worked out. With the melanophores and the xanthophores it is comparatively easy to demonstrate by Wyman's method of nerve cutting that both types of colour cells are under the direct control of nerves. Not only is this true, but as the observations by Fries (1927) have shown, the two sets of chromatophores are excited to action by independent systems of nerves. When *Fundulus* is placed on a yellow background, the xanthophores expand and the melanophores contract, and when it is placed on a blue background, the xanthophores contract and the melanophores expand. These changes do not take place in a denervated portion of the tail fin; hence they are not the direct effect of the environment nor of substances carried in the blood. It is difficult to explain them except on the assumption that each set of chromatophores has its own independent nerve supply.

The maximum number of kinds of chromatophores in fishes cannot at present be stated. Certain flat-fishes possess so far as is at present known the greatest variety of these cells. In *Ancylopsetta* and in *Paralichthys*, as studied by Mast (1916), the individuals will range in tone from almost pure white to near black with all the intermediate stages in grey; and in true colours they will assume a blue, green, yellow, orange, pink, or brown tint in accordance with the background. The physiological details of these remarkable changes are not yet known. Kuntz (1917), who has studied the histology of the skin of *Paralichthys*, finds there only two types of colour cells, melanophores and xanthophores. Such a provision is obviously insufficient. Hewer (1926), who has worked on the skin of the related European dab, recognized in this fish four main types of colour cells: melanophores, erythrophores, xanthophores, and iridophores. This is a more nearly adequate supply, though perhaps still insufficient. To what extent these several types of colour cells are subject to direct nerve control and whether each set is provided with its own independent system of nerve fibres are still to be ascertained. It would not be surprising to find on investigation that these various classes of chromatophores in the flat-fishes possessed as much independence in their innervation as has been demonstrated for the melanophores and xanthophores of *Fundulus*. Although from our present knowledge of fishes such a complicated system as this cannot be said to be established, the probability of its actual occurrence is confessedly very high.

In most fishes thus far studied the colour changes affect the whole body of the animal at essentially the same time and uniformly. Most fishes under change of environment become either light or dark without showing local independent responses in any part of the skin. Many fishes possess a background pattern in their integuments and with the change in tint this pattern becomes more or less distinct, but in most such instances the changes run through a constant series of stages the details of which never show individual deviations. In a number of flat-fishes, however, as demonstrated by Sumner (1911) and by Mast (1916), these animals are coarsely spotted on a coarsely mottled background and finely spotted on a finely mottled one. These types of response are repeated by the fishes when they are placed on artificial backgrounds of checker-board pattern with larger or smaller squares (Fig. 6). As might be inferred from what has already been said, these responses are dependent upon the integrity of at least one eye and disappear wholly when the fish is completely blinded. Although in bringing about these changes it is necessary that the fish should see the background, it is not necessary that at the same time it should see the surface of its own body. A flat-fish resting on a sandy bottom of particular texture will reproduce the details of that texture in its skin even when, with the exception of its eyes, its body is covered with sand. These remarkable pattern changes are therefore not of a kind in which the fish may be said to copy wilfully the condition of the bottom but are obviously reflex in character like local blushing in man. Such observations show that the nervous control of the chromatophores in some fishes at least reaches that

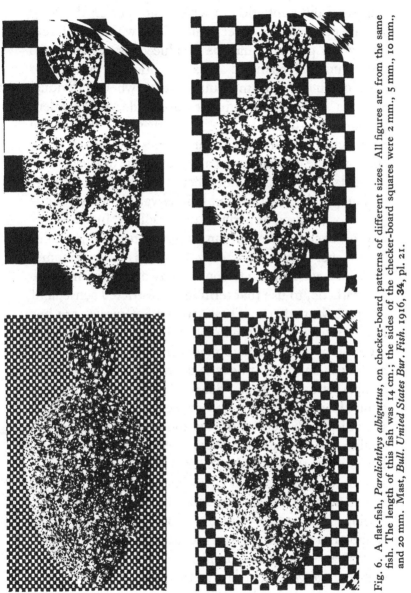

Fig. 6. A flat-fish, *Paralichthys albiguttus*, on checker-board patterns of different sizes. All figures are from the same fish. The length of this fish was 14 cm.; the sides of the checker-board squares were 2 mm., 5 mm., 10 mm., and 20 mm. Mast, *Bull. United States Bur. Fish.* 1916, **34**, pl. 21.

degree of differentiation in which one spot on the body is independent of others and thus wholly different coloration patterns may be produced as responses to environmental changes. This condition is parallel to what obtains in most muscular systems where individual muscles or groups of muscles are under independent nervous control as contrasted with that type of response in which the total musculature comes into action as in the complete contraction of a sea-anemone or the general muscular spasm of a human being in tetanus poisoning.

The problem of the double innervation of chromatophores was raised as early as 1875 by Bert, but even today no decisive answer can be given to this question. In the nervous system of vertebrates the autonomic division is well known to consist of two sets of fibres, the sympathetic, to use that term in its restricted sense, and the parasympathetic. These two sets of nerves are believed to be as a rule physiologically antagonistic, and in the case of the chromatophores the sympathetic fibres have been supposed to excite contraction and the parasympathetic expansion. Spaeth (1916) some time ago pointed out the similarity between melanophores and smooth-muscle cells and thus indirectly suggested the double innervation of these colour cells. He and Barbour (1917) somewhat later published results from a study of the responses of melanophores to ergotoxin and to adrenalin that were indicative of the same conclusion. But it has remained for Giersberg (1930a, 1930b) and for Smith (1931) to attempt a direct experimental attack on this problem. Giersberg maintains that in *Phoxinus*, contrary to the opinion held by von Frisch, only the melanophores as contrasted with the xanthophores and the erythrophores are innervated. Under the influence of such drugs as ergotamin for the sympathetic fibres and pilocarpin, cholin and other such reagents for the parasympathetic, the responses of the melanophores in this fish are such as to give evidence of a double set of fibres. Similar conclusions are drawn by Smith (1931) from his study of the effects of autonomic drugs on the melanophores of *Fundulus*. While these results should be checked from the standpoint of direct effects upon the colour cells concerned, the evidence thus brought forward does point distinctly in favour of double innervation, a conclusion which if once established indicates a still more complicated system of innervation than that based on the independent supply of nerves to the various types of chromatophores.

Although there is abundant and satisfactory evidence that chromatophores in fishes are under nervous control, it is equally true that they can be influenced by hormones, that is, they may be subject to humoral influence. This can be easily demonstrated in fishes prepared according to Wyman's procedure. If a portion of the tail of *Fundulus* is denervated by this process, the melanophores, as already stated, will quickly assume the stellate condition and will remain in this state indefinitely. If now at this stage adrenalin is injected into the fish, the melanophores of the denervated area as well as those of the rest of the body will contract to minute dots, the punctate stage, showing that adrenalin, whatever else it may do, affects the melanophores of this fish directly. If a *Fundulus* whose tail is partly denervated as already described is etherized, the melanophores in the denervated region expand, as do those of the rest of the body, thus showing that ether also acts directly on these cells (Wyman, 1922). The denervated xanthophores of *Fundulus* can also be made to expand by ether. These and numerous other like instances show that the chromatophores of fishes are subject to humoral influences as well as to nerve control. Notwithstanding this fact, however, the great majority of investigators conclude that the chromatophores, and particularly the melanophores, of fishes are controlled primarily by nerves, a means which would appear to be the only one consistent with such remarkable changes of pattern as are seen in the flat-fishes.

The colour changes in amphibians were first observed a little over two hundred years ago by Vallisnieri, who in 1715 described for the first time the changes in tint shown by the common European frog. Subsequently a number of other amphibians were studied with the result that this type of change was found to occur in most frogs, toads, and salamanders. When frogs are kept in illuminated dark-walled chambers they commonly assume a dark tint and when they are retained in light-walled chambers they become pale. These conditions, in conformity with the prevailing earlier opinions, were interpreted as the result of nerve control over the frog's integumentary melanophores. It must be confessed, however, that a perusal of the older papers which deal with this subject and in which experimental procedure such as nerve cutting was indulged in, including the very careful work of Biedermann (1892), leaves on the mind of the reader the impression of complete inadequacy of the evidence.

In 1898 Corona and Moroni showed that when

adrenalin was introduced into the circulation of the frog, the animal became light coloured. This phenomenon was rediscovered in 1906 by Lieben, who investigated the subject with some fullness and showed that in the frog adrenalin by direct action induced a contraction of the melanophores. These observations led a number of investigators to test not only adrenalin but other internal secretions as a means of exciting chromatophoral changes. Chief among such workers were Hogben and his associates, who demonstrated with progressive clearness that in the frog the chromatophores, and particularly the melanophores, were controlled by internal secretions rather than by nerves. The question of a possible nerve control for the melanophores of this animal was carefully reviewed by Kropp (1927), who demonstrated that this factor is at best of very limited application in the frog, a conclusion quite in agreement with the recent work of Slome and Hogben (1928) on the toad *Xenopus*.

In the course of studies on the metamorphoses of tadpoles carried out by Adler (1914), by Smith (1916), and by Allen (1917), it was pointed out that hypophysectomized larvae were always very light coloured. The significance of this change from the standpoint of internal secretions and the melanophore system was first clearly indicated by Atwell in 1919. Two years later Swingle (1921) observed that tadpoles became dark coloured when a part of the pituitary gland was transplanted into them. Hogben and his collaborators, especially Winton, were thereby led to an extensive and thorough investigation of the pituitary gland and its secretions in relation to the chromatophore changes in these animals with results that were truly revolutionary.

The outcome of these investigations has been summarized by Hogben (1924) in his volume entitled *The Pigmentary Effector System*. The complete removal of the pituitary gland from a number of amphibians always leaves these animals in a permanently pale condition even though they are kept in an environment that, under normal circumstances, would induce them to turn dark. An examination of the skin of these pale individuals showed that the melanophores were in maximum contraction. The injection of pituitary extract into such animals was followed by darkening due to the expansion of the melanophores, but this darkening was temporary, for sooner or later such individuals regained their pallor. These results, when coupled with the negative outcome of experimental nerve cutting, led Hogben to conclude that in amphibians the colour changes do not involve direct innervation but depend upon fluctuating amounts of pituitary secretion. The possibility of a dual endocrinal control has been suggested very recently by Hogben and Slome (1931) in their discussion on the conditions presented by the South African clawed toad *Xenopus*.

If the colour cells of the amphibians are controlled through humoral agencies rather than through the direct action of nerves, it follows that such animals are not likely to exhibit pattern changes such as the flat-fishes do, but that they would be limited to changes that would affect each individual as a whole. Tests by Parker (1930) on the tree toad *Hyla* showed this to be the case. If a tree toad is placed on a black and white checker-board background in which the squares are of appropriate size for the animal, the skin, which has a pronounced pattern of its own, will merely change with the illumination from ashen grey to almost black and back again without, however, altering its pattern. There is not the least tendency to reproduce in the skin of this toad the details of the background. Such a result is quite consistent with the belief that the melanophores of these animals are under a humoral control.

Frogs, like fishes, become pallid in a light-coloured environment and dark in a dark-coloured one and lose much of this capacity when they are blinded. That they do not lose it all is probably due to the fact that their skins contain photoreceptors that may in some measure supplement their eyes. These facts suggest very clearly that the initial steps of the colour changes in amphibians take place in their photoreceptors and are nervous. But, as the earlier part of this account shows, these nervous operations must excite secretory processes and the substances thus produced and poured into the blood become the effective elements in calling forth the colour change. To this conclusion Hogben (1924) adds the further statement that if in these animals there is a nervous mechanism for regulating colour control it certainly plays no significant part in their normal colour responses. In this respect amphibians seem to be the reverse of fishes in that the direct nerve control of the fish chromatophore is replaced by a humoral one in the amphibian.

Colour changes in reptiles are limited almost entirely to the group of lizards. Chief among these is the African chameleon, whose remarkable capacity in this respect was recognized by such ancient authors as Aristotle and Pliny. The colour changes of this animal were also made the object of a special investigation by Brücke, who in 1852 published a monograph on this subject that put this whole field of research on a new plane. Since that time a number of other workers have investigated various representatives of the group of reptiles, especially the lizards, with the result that a growing body of information on the colour changes in these animals has accumulated.

Notwithstanding the considerable range of tints shown by lizards in their colour changes, the skins of these animals apparently possess relatively few motile elements. In fact melanophores, which by contracting and expanding expose to view or cover up certain immobile coloured masses, seem to be the chief if not the only means of inducing chromatic changes (Fig. 7). When one recalls the great variety of shades of colour and modifications in pattern that have been ascribed to the chameleon one doubts if so simple a mechanism as that just mentioned is sufficient. Without question many points concerning these operations remain to be discovered, but however much may still be brought to light, it is quite certain from what is now known that the chief factor in the colour changes of these animals is the activity of the melanophore. The expansion and contraction of the pigment in this type of cell is a means whereby the tint of the animal may be darkened or lightened and, as already stated, other masses of highly coloured pigment, probably for the most part immobile, may be variously exposed to view or covered up.

Are the melanophores in reptiles under the direct control of nerves as in fishes or under humoral control as in amphibians? This question has been experimentally investigated by only a few workers. In 1918 Redfield reported work on the melanophore system of the horned "toad" *Phrynosoma*. On stimulating the mouth of one of these lizards for some five minutes with a weak faradic current the melanophores of the whole body contracted

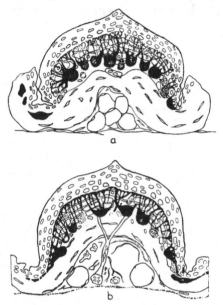

Fig. 7. Chromatophores in the scales of the lizard *Anolis*: *a*, melanophores contracted, animal green in colour; *b*, melanophores expanded, animal dark brown. From von Geldern, *Proc. California Acad. Sci.* 1921, ser. 4, **10**, pl. 9, figs. 11, 12.

and the animal became pale. The spinal cord was then transected under ether at the thirteenth vertebra. On repeating oral stimulation the melanophores, previously expanded, contracted over the whole body as in the previous test. The animal was again etherized and the adrenal glands removed. On stimulating the mouth again the expanded melanophores contracted and the animal became pale from the head back to the region of the cut. The melanophores of the posterior part of the body remained fully expanded. Redfield concluded from this and other like experiments that the melanophores of *Phrynosoma* are under double control, a humoral one, dependent upon adrenalin or some other

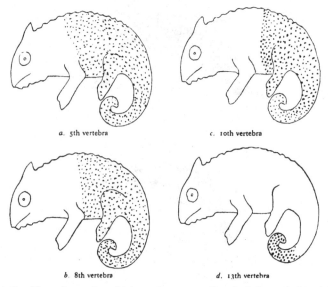

a. 5th vertebra   c. 10th vertebra

b. 8th vertebra   d. 13th vertebra

Fig. 8. Chameleons in which excitement pallor had been induced by stimulation of the mouth and interfered with by severing the spinal cord at different levels. The levels at which the cord was cut are given for each diagram. The dotted areas indicate the regions of skin that remained dark. Hogben and Mirvish, *British Jour. Exp. Biol.* 1928, **5**, 299.

similar substance, and a direct nervous control. In work recently done on the African chameleon by Hogben and Mirvish (1928 *a*, 1928 *b*) it has been conclusively shown that the melanophores of this animal, like those in *Phrynosoma*, are under direct nerve control (Fig. 8), but, contrary to what was demonstrated in *Phrynosoma* by Redfield, there was no evidence in the chameleon of a humoral influence. Nevertheless, when adrenalin was injected into the chameleon, it, like *Phrynosoma*, assumed a marked pallor. Both these pieces of investigation show satisfactory evidence for these two reptiles of direct nerve control of melanophores similar to that already established for fishes, and in one of these there is much that is indicative of a supplementary humoral influence.

Additional confirmatory evidence in favour of the direct nerve control of the melanophores in reptiles has been advanced by May (1924) for *Anolis*. This small lizard exhibits a marked colour change ranging from a very dark brown to a bright green. If a sizable piece of its skin rich in melanophores is transplanted from one spot to another on its body, the transplant will take on a green tint and will cease to show the usual colour changes which are manifested in the rest of the animal's skin. These changes, however, will return to the transplanted piece in about a month, an interval long enough to allow for the growth by regeneration of the subjacent nerves into the transplant. May's results, therefore, harmonize with those of Redfield and of Hogben and Mirvish on the innervation of reptilian melanophores. Hence it seems clear, though the evidence is not very extensive, that the melanophores in reptiles, like those in fishes, are under the direct control of nerves. That they may also be under a supplemental humoral influence is indicated by the work of Redfield. In general, the chromatophoral system in reptiles is much more like that in fishes than it is like that in amphibians. It is essentially nerve-controlled and exhibits relatively little humoral influence. When this whole field is surveyed it must strike one as anomalous that the melanophores of the three classes of cold-blooded vertebrates should present such clear-cut differences. It is highly probable that this colour system took its origin in the fishes, was transmitted from them to the amphibians, and thence to the reptiles. That it should have been first controlled directly by nerves, then by hormones, and finally again by nerves involves a succession of changes that must arouse scepticism in the mind of any zoologist.

CHAPTER III

CRUSTACEAN CHROMATOPHORES

THE crustaceans that exhibit colour changes are found almost exclusively among the larger forms, the malacostracans, and range from the amphipods and isopods to the decapods. The first crustacean in which this activity was observed was the prawn *Hippolyte* (Kröyer, 1842), and the earliest account of the crustacean chromatophore was drawn from the schizopod *Mysis* (Sars, 1867). The systematic study of colour changes in these animals may be said to have been begun in 1872 by Pouchet, and the subject was put on a firm experimental basis by Keeble and Gamble in a succession of monographs issued during the first decade of the present century.

The crustacean chromatophore is a group of closely associated cells or perhaps better a syncitium containing a number of nuclei. The central mass is often so divided as to be suggestive of separate cells. In the expanded condition many long branching processes reach out from the central mass into the adjacent tissue. In the contracted state these processes are said not to be visible. The central mass of the chromatophore contains a dense accumulation of pigment which in expansion passes out in part into the branched processes. This pigment may be of one or more kinds. In *Crangon*, according to Koller (1927), four classes can be distinguished: dark sepia-brown, white, yellow, and red. The brown is most frequently met with and the red least so. In the

movements associated with colour changes, the brown
and white are more active than the yellow and red. The
chromatophores are either monochromatic, in which
case they are always brown; or polychromatic, brown
with any one, two, or three other colours. Thus, brown
is found in all chromatophores, whereas the other colours
occur only in combinations and less frequently. Ordinarily each branch of a chromatophore has its own
colour, though combinations may also occur in the
branches. In *Palaemonetes*, according to Perkins (1928),
there are two classes of chromatophores, one with red
and yellow pigment, and the other containing a substance that is pale yellow by reflected light and slate grey
by transmitted light.

When vigorous specimens of *Palaemonetes* are put into
a dish of sea water with black walls, the red-yellow
chromatophores become fully expanded in about two
hours, the animals assuming in consequence a dark
coloration. When dark *Palaemonetes* are put into a white
dish, the reverse process takes place and the animals
become light (Figs. 9 and 10). These changes are, in
general, recognized for such crustaceans as show chromatic adjustments. The blanching of *Palaemonetes* is,
however, somewhat peculiar. Two or three minutes after
a dark animal has been put into a light dish its chromatophores become surrounded by a bluish cloud which
sooner or later permeates the surrounding tissue. This
blue coloration gradually increases for about an hour
after which, with the contraction of the red and yellow
pigments, it gradually vanishes. This phenomenon was
observed in decapods by Pouchet as early as 1872. In
certain crustaceans it is apparently a regular accompani-

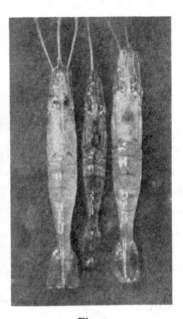

Fig. 9

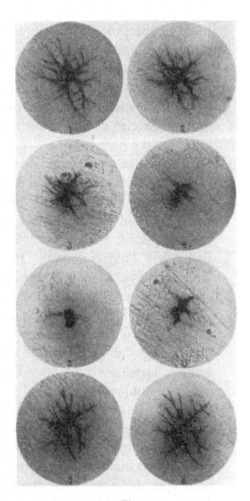

Fig. 10

Fig. 9. Three shrimps (*Palaemonetes*) showing colour differences: left, light coloured from white background; middle, dark coloured, chromatophores expanded as a result of blinding; right, light coloured from injection of extract of eye-stalk from white-adapted shrimp. Perkins, *Jour. Exp. Zool.* 1928, **50**, pl. 3, fig. 21.

Fig. 10. A single chromatophore from the shrimp *Palaemonetes* showing the changes of form from full expansion (1) to full contraction (5) and back to full expansion (8). Perkins, *Jour. Exp. Zool.* 1928, **50**, 101, pl. 1.

ment of the process of blanching. It suggests the disintegration and solution of pigment.

The second type of chromatophore noticed by Perkins in *Palaemonetes*, that containing the pale yellow pigment, is said to respond to the surrounding illumination in a way the reverse of that of the red and yellow chromatophores. In a dark environment the light yellow cells are contracted and in a white one they are expanded. They are, however, so few in number in this shrimp that they have little or no effect on its general coloration. Chromatophores of this type have already been noticed in crustaceans by Pouchet (1876), Degner (1912a), and Bauer and Degner (1913). In *Crangon*, according to Koller (1927), a more complicated form of reaction occurs. This crustacean is not only dark on a dark background and light on a light one, but it assumes an appropriate tint for a yellow, orange, or red background, thus giving indisputable evidence of a true colour response. These several tints are produced by the movements of appropriate types of pigment.

Pouchet (1872) long ago showed that the colour changes in crustaceans were dependent upon the animals' eyes. If one eye is covered or removed, the other eye is sufficient to maintain the normal colour change, but if both eyes are obstructed the changes cease. Only in *Hyperia* has Schlieper (1926), following the work of Lehmann (1923), been able recently to show that tactile organs are probably of much importance in initiating colour changes. In general, however, almost all later workers (Menke, 1911; Megušar, 1912; Degner, 1912a, 1912b) agree in assigning to the eye an exclusive rôle in this respect, a conclusion supported by the results of

Koller (1925) on *Crangon* and other shrimps, and of Perkins (1928) on *Palaemonetes*. When one eye is removed from *Palaemonetes* there is, as might be expected, no marked effect on the chromatic responses, but if both eyes are cut off a complete expansion of the chromatophores follows, and this state persists irrespective of the illumination or the background (Fig. 9). The expansion thus produced is complete in about two hours after the removal of the eyes, the normal period for such a change, and the state ordinarily lasts till the animal's death.

Notwithstanding the opinion held by most early workers such as Keeble and Gamble (1900), and Bauer (1905), external influences appear to have little or no direct effect on the pigment migration in crustacean chromatophores. Taite (1910) observed no effect from light directly applied to the chromatophores of *Ligia*. Koller (1927) showed that heat or ultra-violet radiation had no influence on the migration in *Crangon*, but Smith (1930) has very recently argued in favour of a direct effect from temperature differences on the chromatophores of *Macrobrachium*. Perkins (1928) could detect no change from the direct application of light to the chromatophores of blinded *Palaemonetes*. Practically all investigators, however, agree that the light that enters the eye of the crustacean is the effective means of inducing colour changes. As already shown, this conclusion is well supported by experimental evidence. In consequence of this relation, it was generally assumed by the older workers that in crustaceans nerve tracts must pass from the eyes through the central nervous organs to the chromatophores, and there end in some form of nerve terminal. It is a remarkable circumstance that notwithstanding the fact that many investigators have sought for these nervous connections, no one has been wholly successful in finding them. Retzius (1890) described on the chromatophores of *Palaemon* what he believed to be nerve terminals. His findings, however, have not been confirmed by others, nor has anyone shown in an indisputable way, as has been done for the chromatophores of cephalopods and fishes, that nerve fibres connect with these effectors.

To demonstrate that nerves are essential to the normal action of the crustacean chromatophore, Pouchet, as early as 1876, resorted to nerve cutting. His attempts were followed by those of Mayer (1879), Fröhlich (1910), and Degner (1912*a*, 1912*b*), but to no avail. This method was also attempted by Perkins (1928), who showed that when the ventral ganglionic chain of *Palaemonetes* was severed in the region between thorax and abdomen, no observable effect could be noticed in the subsequent colour changes of the posterior part of the body. After the operation these changes continued as in a normal shrimp. Moreover, if in different shrimps various cuts are made across the abdomen near its connection with the thorax and in such ways that the cuts collectively would sever the abdomen from the thorax, none of these disturbs the colour change in the shrimp except such as pass through the dorsal blood-vessel. Whenever this vessel is severed, the chromatophores, if they are not already expanded, pass quickly into that state and remain so indefinitely. If a small side branch from the dorsal vessel is cut, the chromatophores of the region supplied by that branch become expanded and remain permanently so, while those on the rest of the abdomen continue to show the characteristic changes. From experiments of this kind it was suspected that chromatophoral nerves may accompany the blood-vessels and be distributed to these effectors along the lines of the dorsal blood-vessel and its branches. When, however, these vessels were studied histologically, not the least sign of nerve fibres could be discovered on them, and Perkins was forced to the conclusion that the agent controlling chromatophore action in *Palaemonetes* was some constituent of the blood itself.

In 1925 Koller showed that if the blood from a dark *Crangon* is drawn and injected into a light one, the light shrimp quickly becomes dark. Perkins (1928) was unable to carry out successful experiments of this kind on *Palaemonetes*, but attempts were made to discover in the body of this shrimp the source of a substance that might induce such a change. It is evident from the experimental results already brought forward that the light entering the eye of *Palaemonetes* is the outward stimulus to colour change, and that what immediately induces the change in the chromatophore is a substance carried in the blood. Where is this substance produced? Perkins tried watery extracts from many organs in the body of *Palaemonetes*, but without success. Finally he took the eye-stalks from several light-coloured *Palaemonetes*, crushed them in sea water, and injected the extract thus obtained into a blinded *Palaemonetes*, whose dark chromatophores, in consequence of the treatment it had undergone, were expanded. Within an hour the dark pigment in these chromatophores had contracted and it remained so for about a day, after which it again expanded (Fig. 9). No such changes were produced by the injection of sea water into the shrimp, and it was therefore concluded that when the retina of *Palaemonetes* is stimulated by the light from a white background, the eye-stalks of this shrimp produce a substance that passes into the blood and excites a contraction of the dark pigment of those chromatophores to which the blood is distributed. As confirmatory of this view, Perkins succeeded by a very simple operation in closing temporarily the dorsal abdominal blood-vessel. When in a light-coloured *Palaemonetes* the blood-vessel was closed for an interval, the portion of the shrimp behind the

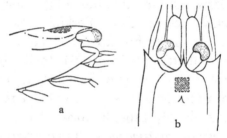

Fig. 11. Cephalothorax of the shrimp *Crangon*: *a*, seen from the side; *b*, seen from above. The areas lined obliquely show the rostral region in which the "black organ" is located. Koller, *Zeit. vergl. Physiol.* 1928, **8**, 604.

region of closure became dark; on releasing the vessel again so that the current of blood was re-established, this region returned to its original light state. Although it was carefully sought for, no substance that would produce the reverse effect on the dark pigment in the chromatophores of *Palaemonetes* was found by Perkins.

These results were abundantly confirmed on *Crangon* and on *Leander* by Koller (1928), and on *Macrobrachium* by Smith (1930). Koller also showed that if an extract is made from the rostral region of *Crangon* (Fig. 11), this extract, when injected into a shrimp the dark pigment of whose chromatophores has contracted, will cause this pigment to expand. Thus, a so-called black organ was located in *Crangon* whose function in the production of hormones was the reverse of that in the eye-stalk. Koller designated the substance produced by the eye-stalk as "contractin" in consequence of its action on the dark pigment of the chromatophores, and that produced by the black organ as "expantin". It is interesting to note that the black organ occurs in that part of the adult *Crangon* where the larval nauplius eye has disappeared. Perhaps some of the tissue of this primitive visual organ persists in the adult shrimp at this spot, and secretes the hormone for chromatophoral expansion. Thus, both contraction and expansion of the chromatophores in *Crangon* may be induced by hormones, contraction by the eye-stalk hormone and expansion by the rostral hormone.

That the internal secretions responsible for many of the colour changes in shrimps have many of the general characteristics of the hormones from vertebrates has been abundantly shown by Koller (1928, 1929, 1930). Thus, the effectiveness of neither the extract from the eye-stalk nor that from the rostral organ is changed by boiling. Further, this effectiveness is resistant to digestive operations. The secreted substances, moreover, remain active even after excessive dilution, such as one part in a hundred thousand. Finally, these substances are commonly not specific, but those from one species will call forth characteristic responses in another. In fact, Koller and Meyer (1930) have shown very recently that extracts from the eye-stalks of *Crangon* and of *Praunus* will induce contractions in the melanophores of the fishes *Gobius* and *Pleuronectes*; in other words hormones from invertebrates may affect the melanophores of vertebrates. The substances under consideration, therefore, exhibit all the important qualities ordinarily ascribed to hormones and though they are unknown chemically it seems fairly clear from their properties that they do not fall under the class of ferments.

If the colour changes in crustaceans are controlled by hormones, it is not surprising that all the experiments on the cutting of nerves with the view of discovering the tracts over which the effector impulses pass should have resulted negatively. Nor is it remarkable that no one has found nerve terminals for crustacean chromatophores. There is apparently no good reason to suppose that chromatophoral fibres or terminals exist in crustaceans. In this respect the crustacean chromatophores are like those in amphibians and have been contrasted with the chromatophores in the fishes and the reptiles. The fish and even the reptile chromatic organs are commonly believed to be completely under nerve control, while those in the crustaceans are apparently purely humoral and without nervous connections.

Although the humoral control of the crustacean chromatophores may seem to be the real solution of this particular aspect of the general problem, it must be kept in mind that the implications of this solution are by no means simple. When the details of these relations in such an animal as *Crangon* are pictured, the complexity of the situation must be apparent. This shrimp, according to Koller, adapts itself well to white, black, yellow, orange, or red backgrounds. Are we to assume a separate hormone or possibly a pair of hormones for each of these changes? What induces at the same moment, and in the same chromatophore, the outward migration of the dark pigment and the inward migration of the light pigment (Koller, 1928)? Although it is easy to understand how the "light" organ, located as it is in the eye-stalk of the shrimp, may be excited to action from the eye, in what way is the "black" or rostral organ brought into action? Are there nerve connections between this and the eye or is the relation here also humoral? Questions such as these, for which we have at present no adequate answers, show that we are as yet far from a complete understanding of the operations of even such simple chromatophores as those in the crustaceans, and suggest a degree of complexity in the functions involved that strain to the limit any explanation thus far offered for the activities of these parts (Piéron, 1913, 1914).

The changes in the crustacean chromatophores are of special interest because of the significance they have for the process of chromatophoral reactions in general. Although the responses of the chromatophores in shrimps and other like forms have been declared to be humoral, they are, nevertheless, known to be dependent upon the animals' eyes. In these organs, as a result of increased illumination, for instance, nervous activities are set up which so excite certain parts of the eye-stalks that there is secreted from these parts into the blood a substance which when it reaches the chromatophores induces them to contract. Thus, the activation of this type of chromatophore is not purely humoral but also involves the eye. The process is evidently a double one, first the nervous response of the optic apparatus and next the secretory action of some hormone-producing organ in the eye-stalk. The process therefore involves both nerve and gland and may in consequence be described, to use a term already employed by Fredericq (1927), as neuro-humoral. Such a combination of double elements leading to excitation seems to be characteristic of crustacean chromatophores in general.

That this method of activation is applicable to vertebrate chromatophores may seem at first sight quite remote, but a thoroughgoing consideration of such a view shows it to be not altogether impossible. In fact, much may be said in favour of a comparison between the conditions of chromatophoral excitation in crustaceans and in amphibians. In amphibians the chromatophores, in accordance with the work of Hogben and his associates, are controlled through the secretion of pituitrin. But in these animals, as in the crustaceans, the eye is essential to the secretion of the controlling substance. In some way the activities of the eye induce in the amphibian pituitary gland, in all probability in its pars intermedia according to the recent work of Allen (1930), the secretion of a hormone which after transportation by the blood excites expansion in the melanophores of the animal. The pituitary gland of the amphibians acts then as the rostral organ of *Crangon* does in that it is excited to secretion through impulses that start in the eye. Thus, in this respect the crustaceans and amphibians may be shown to be almost exactly parallel, and if the control of crustacean chromatophores is described as neuro-humoral, that of the amphibian colour cells may likewise be so designated.

But notwithstanding this general agreement in these two groups of animals certain differences in details appear to be present. Thus, in most amphibians Hogben has found only one effective hormone, pituitrin, which, as already pointed out, induces chromatophoral expansion. Contraction of amphibian chromatophores is believed to be a return to the resting state. In the crustacean *Palaemonetes* Perkins has likewise found only one hormone but this induces contraction, not expansion; thus, *Palaemonetes* presents a condition the reverse of that in amphibians. Whether on further study both amphibians and *Palaemonetes* will be found to possess additional and complementary hormones is still to be determined. If such proves to be the case, they will then agree exactly with what Koller has shown to be true for *Crangon* and several other shrimps, namely, in one and the same animal there are both contracting and expanding hormones. That there are in amphibians other organs concerned with colour control than the hypophysis has been shown by Kropp (1929), who found that an extract from the eye of a tadpole with expanded melanophores would, when injected into one with contracted colour cells, cause these cells to expand, a reaction which, however, has the same direction as that induced by pituitrin. But these various peculiarities are all matters of detail and do not in any way militate against the general conclusion that in amphibians as in crustaceans the chromatophoral control may be truly designated as neuro-humoral.

Whether such a conception can be applied to the fishes and the reptiles is a matter of uncertainty. The control of the chromatophores in these animals, as already stated, is commonly attributed to nerves, and is so different in principle from that in the crustaceans and amphibians that any unification seems impossible. In both fishes and reptiles, however, the colour cells are extremely responsive to a number of injected substances, such as adrenalin, pituitrin, and the like, all humoral in nature. So far as I know in all fishes and reptiles thus far tested adrenalin induces directly a contraction of the melanophores, and several other substances might be mentioned in which a direct action on the colour cells of these animals can be shown. If these chromatophores are thus open to humoral action, may it not be possible that the so-called nerve control of them is due to hormones secreted by the terminals of the chromatophoral fibres in such close proximity to the colour cells as to excite them to respond? Such a condition is by no means impossible and if it proved to be an actuality, it would bring the method of activation in the fishes and reptiles into reasonable harmony with that in the crustaceans and amphibians. In these forms the nervous activity excited by changes in the environmental illumination induces a secretion in some well circumscribed distant gland, whence the secretion is carried by the blood to the responding chromatophores. In the fishes and reptiles, according to the present assumption, the optic impulses, after passing through the central apparatus, would excite secretory activities in the efferent nerve terminals close to the chromatophores, and the substances thus secreted would need only to pass across the very narrow lymph spaces between the terminals and the chromatophores in order to excite this colour organ to action. The chief difference between the method ascribed to crustaceans and amphibians and that for fishes and reptiles is that in the former hormones are secreted at some distance from the chromatophores and by special glands and in the latter they are assumed to be secreted very close to the responding cells and by the nerve terminals themselves (Fig. 12). Such a conception as this would unify completely the general scheme of chromatophoral action in both crustaceans and in vertebrates and would do away with that strange anomaly

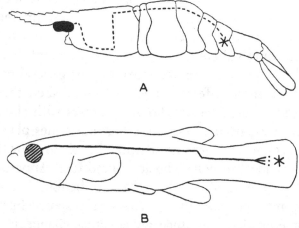

Fig. 12. Diagrams of the neuro-humoral systems in a shrimp (*A*) and in a fish (*B*). In the shrimp the neural part is entirely within the eye-stalk (black) whence the humoral part (dotted) is represented by the course of the hormone carried by the blood from the eye-stalk to the chromatophore. In the fish the neural part is represented by the eye, central tracts, and nerves (black) to the immediate vicinity of the chromatophore where the hormone is secreted by the nerve terminals. From the nerve terminal to the chromatophore is the humoral part (dots).

already pointed out in which the amphibians are set in contrast with the fishes and the reptiles. According to this hypothesis the two so-called types of control, nervous and humoral, occur in each instance, and confusion has arisen because in some cases one type has been described to the exclusion of the other even though both were present. Such a unified view has already been suggested by Giersberg (1930c) and much can be said in its favour.

The correctness of the neuro-humoral hypothesis as applied to crustaceans and amphibians is beyond doubt. Its application to fishes and reptiles remains to be worked out. So far as I know, few if any observations have been made which bear on this question. It will be recalled that in the discussion in the preceding chapter of the two assumed methods of control it was pointed out that nerve control admitted of local activities and differences whereby changes in colour pattern could be excited, whereas humoral control, since it was accomplished by the blood as a whole, must necessarily be general rather than local in its effects. It is well established that in flat-fishes local chromatophoral responses with changes in pattern do occur. Assuming that hormones play the part in colour changes just attributed to them, how then can these local responses be accounted for? Apparently the only solution of this difficulty is to be found in local secretion strong enough to excite local response but not strong enough to produce an effective change in the body as a whole. In other words, the locally secreted hormone would be just sufficiently concentrated in the region of its production to affect the local chromato-phores and would lose this capacity for more distantly placed colour cells when through diffusion it became diluted. Such an assumption would account for local action and at the same time allow the colour changes to remain at that stage on a humoral basis. That this explanation is probably correct is seen from the recent observations by Meyer (1930), whose experiments on flat-fishes bear on this very point. According to this investigator, serum from a flat-fish that had been made dark by one to two weeks residence on a black background was injected into a light flat-fish with the result that after about five minutes the region of injection became locally dark. The local darkening increased for some half an hour after which it disappeared. The reverse experiment in which serum from a light fish was injected into a dark individual was followed by a corresponding result, namely, the development of a temporary local light spot in the region of injection. As serum from a dark fish was followed by no appreciable change when it was injected into another dark individual and no change occurred on the injection of "light" serum into a light fish, it was concluded by Meyer that the temporary colour changes were true humoral changes and not operative results. These observations are in perfect accord with the assumption of local secretion and show in a most indubitable manner that even in the flat-fishes where local chromatophoral responses are better developed than in any other animals, a humoral element is quite clearly present and represents in all probability the last of the steps in the mechanism for colour change.

That nervous tissue should be endowed with secretory function is a somewhat unusual conception. Nevertheless, over a decade ago Speidel (1919, 1922) described groups of large nerve cells in the spinal cords of several fishes and concluded from their histological condition that they were concerned with internal secretion. But this interesting suggestion was never followed up physiologically. The adrenal glands, however, afford an instance in which the evidence for nervous secretion is very much more fully attested. These glands in the higher vertebrates consist of two clearly separated portions, the cortex and the medulla. The medulla, which is rich in chromophil cells, is the portion of the gland from which adrenalin is obtained, none of this material being produced by the cortex. These two parts of the gland arise embryologically from very different sources. The cortex is derived from tissue associated with the primitive kidneys. The medulla comes directly from a portion of the developing sympathetic nervous system; in fact, the cells that compose it are modified sympathetic ganglion cells and adrenalin is their secretion product. If adrenalin, which as already indicated is a most potent material in affecting chromatophores, is secreted by the modified sympathetic cells of the adrenal gland, it is not a far reach of the imagination to ascribe to the nerve terminals of sympathetic cells a secretory action whose product may control colour cells. These relations, which are not very different from those advanced by Elliott in 1904, make the idea of secretions from nervous tissue less of a novelty than at first sight it appears to be and support the general hypothesis of the interrelation of nervous elements through secreted substances.

If the variety of chromatophoral responses in crustaceans and in vertebrates is called forth by secretions excited in nearby nerve terminals or more distant glands through visual impulses, such secretions must be varied and relatively numerous. Koller (1928, p. 611), in discussing these relations in the shrimps, as has already been mentioned, distinguished two such products: contractin, concerned with the blanching of the shrimps and derived from their eye-stalks, and expantin, concerned with their darkening and produced by their rostral organ, both hormones in nature. Contractin and expantin, however, are terms that probably designate classes of hormones rather than individual hormones, for the experimental results from various crustaceans, as reported by Koller and Meyer (1930), show such diversity in the reactions of the several species tested as to imply a marked difference in what might be assumed to be one substance. This condition is perhaps more clearly seen in the fish *Fundulus*. In this animal, as has already been pointed out, there are two well-defined groups of chromatophores, melanophores and xanthophores. These, according to the observations of Fries (1927), act independently, for when the fish is on a white background both are contracted, on a dark background both are expanded, on a yellow background the xanthophores are expanded and the melanophores contracted, and on a blue background the xanthophores are contracted and the melanophores expanded. These conditions have already been cited in part to show that melanophores and xanthophores in this fish are independently innervated, but they afford equally sound ground in favour of independent classes of secretions. In fact, it seems impossible to explain these responses except on such an assumption. It is by no means clear that in every animal with active chromatophores an expanding substance and a contracting substance must necessarily both be present. In most amphibians, according to Hogben, only one substance, pituitrin, an expanding agent, is at hand. The resting melanophore is assumed to be the contracted melanophore and the degree of its expansion is believed to be dependent upon the amount of pituitrin temporarily present. In fishes, however, the resting melanophore is approximately midway between full expansion and full contraction, the so-called stellate stage. In such forms, therefore, an expanding hormone and a contracting one appear to be necessary. Thus, for example in *Fundulus*, with its two independent classes of chromatophores, melanophores and xanthophores, it is necessary to assume four humoral substances which, using the terminology already begun by Koller, might be called melanoexpantin, melanocontractin, xanthoexpantin, and xanthocontractin. How many such substances may be present in the chromatophoral systems of various animals simple and complex cannot of course be stated, but enough has been said to show that if the neuro-humoral interpretation of chromatophoral action is correct, as there is every reason to believe it is, a considerable array of such hormones may be anticipated, an array which, however, may not be without its limitations, for, as already pointed out, some crustacean hormones have already been shown by Koller and Meyer (1930) to be effective in vertebrates. These hormones are probably complex organic substances; at least Charles (1931) has shown that no certain correlation can be detected between the calcium and the magnesium content of the serum from *Xenopus* and its pigmentary activities.

From what has been advanced in these two chapters it must be evident that secretions in the nature of hormones play a considerable part in the control of chromatophores and that these secretions are chiefly of nervous origin. Some of them, like adrenalin and pituitrin, come from well-defined nervous organs, the adrenal medulla and the pituitary gland, but others appear to come from the nerve terminals of sympathetic fibres, a state of affairs that suggests that the secretory activity of nerves may be a phenomenon of wide application and of great biological significance.

CHAPTER IV

NEURO-HUMORALISM

Effectors. If the control of chromatophores is to be ascribed to a neuro-humoral activity, to what extent does this principle apply to other effectors? The nettling capsules or urticators of the coelenterates, as already stated, are not known to be under nervous control in any way, but are discharged as the result of direct stimulation from the exterior. Consequently these effectors, which are thus essentially independent, are not to be considered from the standpoint of neuro-humoral excitation. On the other hand, luminous organs, cilia, and electric organs in certain instances at least are under the control of nerves, but their methods of excitation have been investigated to so limited an extent that no significant facts are at hand bearing on the possibility of neuro-humoral stimulation. Glands, however, have been more fully studied in this respect and present conditions which from this standpoint are of no small importance. The pancreas was the first effector in which a humoral control was demonstrated. As is well known, Bayliss and Starling in 1902 showed for the first time that secretion in this gland was not excited by nerves but by a hormone. In this excitation the acid chyme from the stomach acts on the walls of the duodenum and produces a substance which when carried by the blood to the pancreas stimulates this gland to secrete. The active substance produced in this operation is the so-called secretin, the type of all hormones. The usual kind

4

of excitation met with in the pancreas is purely humoral and there is no reason to assume that any form of nervous activity is involved in it. So far as the present discussion is concerned it shows quite indisputably that a gland may be brought to secrete through the direct action of a humoral agent. The salivary glands, on the other hand, have been described since the days of Ludwig as under nervous control. Is it possible that this so-called nervous control is in reality neuro-humoral and that the secretory cells in these glands are excited to action through substances produced by excitatory nerve endings as the chromatophores in fishes and in reptiles are? That the salivary cells are open to humoral stimulation was shown by Florovsky in 1917, who demonstrated an increased flow of saliva when adrenalin was allowed to enter the denervated salivary gland of the cat (Sharpey-Schafer, 1924). That the normal excitation of the gland is in reality neuro-humoral seems probable from observations by Demoor (1913). This investigator collected spittle from a dog whose salivary glands were excited by the electrical stimulation of the chorda tympani. When this spittle was injected into another dog the glands of this second individual became active and discharged saliva, a circumstance difficult to understand except on the assumption that the saliva from the first dog contained a substance produced in its glands by nervous action and capable of exciting the glands in the second dog. Such an assumption implies a neuro-humoral type of stimulation, and raises the presumption in favour of this interpretation for all so-called nervously controlled glands.

Evidence bearing on the question of neuro-humoral

action in muscle comes chiefly from the simpler types of this tissue such as smooth muscle and the muscle of the vertebrate heart. As early as 1908 Howell and Duke showed that the fluid from a perfused frog's heart whose beat had been modified by the stimulation of its vagus nerve was richer in potassium than that from a heart whose beat had not thus been changed. Following this suggestion Loewi re-investigated the subject and in 1921 began the publication of a series of most important papers on the humoral relations of the heart muscle in vertebrates. In the first of these papers (1921), in which the hearts of frogs and of toads were studied, he showed that significant changes took place in the Ringer's fluid contained in a heart in consequence of the stimulation of its vagus nerve. A normally active frog's heart free of blood was filled with Ringer's fluid which, after a given period, was drawn off and set aside. A second equal volume of Ringer's fluid was introduced into the heart and retained there for a like period during which the vagus nerve of the given heart was stimulated. This second volume of fluid was then withdrawn and also set aside. After the heart had returned to normal beat the first fluid was introduced into it and without observable effect. The introduction of the second volume of fluid, however, was followed by an inhibition of the heart beat in all respects like that produced by the stimulation of its own vagus nerve. Thus, the fluid in a heart whose vagus nerve is stimulated receives some substance which enables it when re-introduced into the heart without vagus stimulation to change the beat of that organ pre-cisely as the activity of its own vagus nerve would have done. This substance is therefore rightly described as a vagomimetic material and is in all probability the substance whereby the normal vagus change is accomplished. In a similar way Loewi demonstrated a substance concerned with the acceleration of the heart, and in a series of papers extending over the last decade he has supported with ever increasing certainty his original contention that the nerves that act on the heart do so through secreted substances essentially humoral in nature. He thus laid the foundations for a neuro-humoral interpretation of the control of the heart.

Although some investigators failed to confirm Loewi's results, the majority gave his conclusion unqualified support. Thus, Brinkman and van Dam in 1922 tested Loewi's view in an interesting and novel way. They linked two frogs together so that the perfusing Ringer's fluid from the heart of the first frog was led by a narrow tube into the blood-vessels that supplied the stomach of the second frog (Fig. 13). When the vagus nerve in the first frog was appropriately stimulated the effect on its heart was followed by distinct contractions of the stomach in the second frog. This demonstrated in the opinion of the authors the production of a humoral material by the vagus stimulation of the heart in the first frog, and the transmission of this material to the stomach of the second one with the resultant responses. That this transmission was dependent upon a transfer of substance and not upon shifting fluid pressures was shown by Ten Cate (1924), who repeated the experiment but with the following modifications: midway on the connecting narrow tube between the heart of the first frog and the stomach of the second one a chamber for the relief of pressure was inserted and thus the varying pressure of the heart beat was prevented from reaching the stomach. Under such circumstances the test organ in the second animal responded in a typical way to the effects of vagus stimulation applied to the heart in the first one.

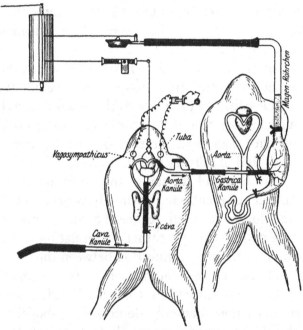

Fig. 13. Plan of an experiment in which Ringer's solution from the heart of one frog whose vago-sympathetic nerve was stimulated is led to the walls of the stomach of a second frog to test for the presence of a vago-humoral material. Brinkman and van Dam, *Arch. ges. Physiol.* 1922, **196**, 75.

Results essentially similar to these have been obtained by numerous investigators not only upon cold-blooded vertebrates but also upon a number of warm-blooded forms such as the rat, rabbit, cat, and dog. Ample surveys of this general field have already been made by Fredericq (1925 and 1927) and by Demoor (1929). From

the earlier studies and from these surveys it is quite clear
that the vertebrate heart muscle is modified in its beat
by substances that are produced by its nerves and that
act directly upon it; in other words the vertebrate heart
is under a neuro-humoral control.

Evidence of a very interesting kind on humoral action
in smooth muscle has very recently been advanced by
Cannon and Bacq (1931), in their experiments on the
cat. If in a cat whose heart has been denervated and
whose adrenals and liver have been excluded from
action, the sympathetic fibres to the smooth muscles of
the tail hairs are stimulated, there will occur for about
two minutes a gradual increase of blood pressure, of
heart rate, and of salivary secretion. If the flow of blood
in the hind legs and the tail of the cat is hindered, the
phenomena just mentioned are much reduced or delayed
until the flow is restored. Since the only tissue known
to be affected by the sympathetic stimulation is smooth
muscle, since the only connection between the hind part
of the cat and the responding denervated organs in its
fore part is the blood stream, and since interference with
this stream either markedly depresses or abolishes the
response, Cannon and Bacq conclude that a substance is
given off from the smooth muscle of the hair follicles
into the blood and carried by it to the distant organs,
which respond as though they were stimulated by sympathetic impulses. To this substance the authors gave
the name of sympathin. Thus, smooth muscle like the
heart muscle may be involved in the production of
material which, hormone in character, may influence
distant parts.

Although most of the evidence for humoral action in

muscle has come from the vertebrate heart or from smooth muscle, there are a few observations to show that vertebrate skeletal muscles may not be without this property. Brinkman and Ruiter in 1924 and 1925 showed that Ringer's fluid that has passed through an active skeletal muscle of a frog was stimulating to the cloaca and rectum of a second frog, a condition indicative of humoral agents originating in cross-striped muscle. Thus, possibly all innervated muscle, cross-striped, cardiac, or smooth, may in activity produce substances that give evidence of humoral function as determined by such test organs as the heart, stomach, intestine, salivary glands, and the like.

The exact source of the several humoral substances thus associated with muscle has not excited much attention. In the most recent contribution to this subject, that by Cannon and Bacq, the smooth muscle cell is continually referred to as the element from which the humoral substance emanates. Such a view is a natural one, though as a matter of fact there appears to be no special evidence in its favour. It seems quite evident from the conditions under which humoral substances in muscles are produced that these substances must originate either from the muscle cells or from the nerve terminals. Because of the small size of the terminals it would be natural to assume that the given substance was produced by the muscle, but if the most recent work on chromatophores has any meaning at all, it points most conclusively to the nerve terminals as the sources of these materials. Although it is true that the terminals are of small size, it must be remembered that the substances under consideration, such for instance as adrenalin, are enormously powerful as shown by the degree of dilution that they may suffer and still remain effective, and that therefore a nerve terminal may be amply large enough to secrete them. Crucial evidence on this point is not at present to be had. Brinkman and Ruiter showed in 1925 that when Ringer's fluid was passed through a muscle that was being stimulated through its nerve and yet was rendered immobile by curari, the outflowing fluid was nevertheless effective in stimulating such an indicator as the intestine. This observation, so far as it goes, favours the nerve rather than the muscle as the source of the humoral material and leads to the conclusion that muscles, like chromato-phores, are stimulated to action by substances that emanate from the nerves with which they are associated. Such being the case, changes in the heart beat, in stomach or intestinal movements, in salivary secretion, or in the movement of the hairs on the cat's tail must then be due to substances produced by nerves and not by muscles, and it is the overflow of these substances into the general circulation that calls forth such changes as the various indicator organs may show, all of which may be said merely to measure this humoral excess. What these nerve-generated substances are remains to be deter-mined. That adrenalin may be one of them is not impossible (Elliott, 1904, 1905), and the remarkable parallel shown by Dale and Gaddum (1930) between the responses of certain muscles as excited by their normal nerve supply and by acetyl choline points to a possibility that this substance is another. But this aspect of the subject has scarcely been touched upon. Little is really known of the way in which effectors under nerve control are actually stimulated, but such as has been ascertained, particularly in relation to chromatophores, glands, and muscles, supports the belief that nerve-controlled effectors are normally excited to action by neuro-humoral means.

Receptors. It is an interesting fact that in 1909 and 1910 Botezat published two papers which dealt with the vertebrate integumentary sense organs and in both of which he advanced on almost purely histological grounds what is essentially a neuro-humoral view of the relation of secondary sense cells to their associated nerve fibres. In discussing the specialized types of tactile cells in vertebrates he expressed the opinion that under mechanical influence these cells might produce a secretion that would stimulate the nerve fibres applied to them, and in a similar way he thought it possible that the normal stimulation of taste cells might induce them to secrete and thereby excite the fibres concerned with the sense of taste. This assumed secretory function of secondary sense cells was emphasized by Botezat in that he proposed to designate all such cells as " Sinnesdrüsen-zellen", a novel and suggestive terminology. Botezat's idea of sensory secretion might well have been extended to the whole epithelial constituent of the vertebrate integument whose cells, activated by appropriate stimuli, might be brought to exude a material which would excite the free-nerve endings in that part of the skin. But no attempt was made to extend this conception to the nervous system in general. So far as it was applied, however, it reproduced completely the neuro-humoral concept. Botezat's hypothesis met with some opposition, was never tested physiologically, and was apparently soon lost to sight. It remains, however, a very early if not the first application of this concept to a limited field in receptor activity.

Although the relations of secondary sense cells to their nerve fibres appear never to have been tested experimentally, those of the fibres to the cells have been worked upon with considerable success. This is particularly true of the vertebrate taste-buds. Each taste-bud, whether upon the outer skin of the vertebrate, as in many fishes, or on the surface of the mouth, as in the higher members of this phylum, consists of a cluster of secondary sense cells supplied with nerve fibres. As early as 1876 Vintschgau and Hönigschmied observed that when the glossopharyngeal nerve in the dog was cut on one side of the head, the taste-buds disappeared from the corresponding side of the tongue. Ranvier in 1882 stated that forty-eight hours after section of this nerve the taste-buds began to show signs of degeneration and that forty days later they had entirely disappeared. In 1887 Griffini published an account of his experiments on rabbits, in which he declared that the first signs of degeneration in the taste-buds were to be seen twenty-three hours after the nerve had been cut and that their complete disappearance had taken place by the twenty-eighth day after the operation. Regeneration of the buds was well advanced seventy-six days from the time of the original cut. These observations were confirmed in all essential respects by Sandmeyer in 1895 and by Meyer in 1896. In 1920 Olmsted began similar work on cold-blooded vertebrates by experimenting on the taste-buds of the barbels of the catfish *Ameiurus*. Here he found that on severing the nerve the buds disappeared completely in from eleven to thirteen days. They reappeared only on the regeneration of the nerve. As the buds developed only after the nerve had regenerated, Olmsted concluded that the nerve was the causative factor in the formation of the bud. May, in 1925, showed that each taste-bud in *Ameiurus* was associated with what were essentially neuro-fibril branches from the nerve fibres proper. These branches were found to degenerate a few

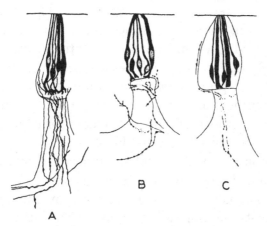

Fig. 14. Taste-buds from the barbels of the catfish *Ameiurus*: *A*, normal taste-bud with innervation; *B*, taste-bud eleven hours after nerve was cut; *C*, taste-bud fifteen hours after nerve was cut; nervous arborization disintegrated. May, *Jour. Exp. Zool.* 1925, **42**, 381.

hours after the nerve had been cut, to be followed in time by the degeneration of the bud (Fig. 14). Only after the regeneration of the neuro-fibril branches does a new bud appear. Thus, May supported with much detail the general conclusion arrived at by Olmsted. May further declared from the evidence at hand that it is probable that some substance in the nature of a hormone is being given out continually by the gustatory fibrils and that this substance is necessary for the existence of the bud.

When by severing the nerve it is excluded from the bud, this structure degenerates, nor does the bud regenerate until by the reformation of the fibrils a new supply of the given substance is brought into the region of the bud.

The lateral-line organs of the lower vertebrates appear to be related to their nerve supply in much the same way as the taste-buds are to theirs. These organs form definite tracts on the head and on the sides of fishes and of certain water-inhabiting amphibians. They are sometimes on the immediate surface of the skin and at other times in grooves or more frequently in canals immediately under the skin. They are probably concerned with the reception of water vibrations of low rate. The tract running along the side of the trunk of a fish, the lateral line proper, is most convenient for experimentation, for it consists ordinarily of a series of bud-like organs commonly in a canal and innervated by the large lateral-line branch of the vagus nerve. This branch, which is usually directly under the skin, can be easily cut near its anterior end, and the effect of its severance on the lateral-line organs can be studied in preparations made from the part of the tract lying posterior to the cut. Each lateral-line organ consists of an open bud-like group of sensory cells very like a gigantic taste-bud and among these cells terminate the ultimate divisions of the lateral-line branch of the vagus nerve (Fig. 15, A).

Brockelbank, who, in 1925, reported on the effect on these organs of cutting the lateral-line nerve in the cat-fish *Ameiurus*, found that within four days of the time when the nerve was cut the lateral-line organs began to degenerate and that this degeneration, far advanced after a little over a week, continued for at least thirty-five days.

Eventually the organs degenerated fully (Fig. 15, B). The regeneration was to be seen at fifty-four days after the initial operation and was practically complete at 116 days. These organs then, like the gustatory buds, degenerate when their nerve degenerates and regenerate later, presumably after it has regenerated. Thus, what little is known of the lateral-line organs and their nerve relations agrees with what has been demonstrated much more

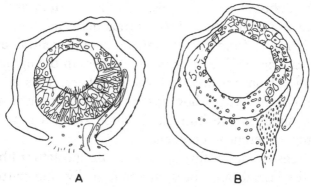

A   B

Fig. 15. Cross-sections of the lateral-line canal of the catfish *Ameiurus* in the region of lateral-line organs: *A*, normal lateral-line organ; *B*, completely degenerated lateral-line organ thirty-five days after the nerve had been cut. Brockelbank, *Jour. Exp. Zool.* 1925, **42**, 304, 305.

fully for the gustatory buds and their nerves, and presumably the same explanation that has been offered for the relation of nerve to cell in the gustatory system applies in the lateral-line system.

In both these instances the influence of the nervous part of the mechanism on the secondary sensory cell is obviously trophic. The continued life of the receptor cells in gustatory or in lateral-line organs is dependent upon the presence of appropriate nerve terminals. It seems quite probable, as suggested by May for the gustatory terminals, that in both instances the nerve terminals secrete substances essential to the integrity of the groups of secondary sensory cells with which they are associated. Such a relation is strictly neuro-humoral and, though it has not been demonstrated experimentally, the conditions just described strongly favour it. Both these examples are good instances of what has often been pointed out, namely, the independence of the normal nerve impulse and the trophic influence that passes over a nerve fibre to keep its sense cells intact. In both these instances the direction of the normal nerve impulse over the fibres is from the peripheral to the central end, whereas the trophic influence must pass in the opposite direction from the centre to the periphery. This difference is sufficient to show that the trophic impulse and the ordinary nerve impulse in these systems must be entirely different phenomena.

If sensory nerve fibres, such as the gustatory fibres and the lateral-line fibres, are continually transmitting trophic impulses peripherally and thereby controlling the life of distally situated groups of sensory cells, may it not be possible that to some extent at least the common cutaneous sensory nerves play a somewhat similar part in maintaining a normal state in the ordinary epidermal layers of the skin? It is, of course, quite obvious that whatever relation there may be between the cutaneous nerves and the epidermis, this relation is by no means so intimate as in the gustatory and lateral-line systems. But that such a relation may subsist is indicated by certain diseases originally supposed to be integumentary as, for instance, *herpes zoster* or shingles. This disease in its obvious manifestation appears in man chiefly on the thoracic skin in the form of transverse bands of acute inflammation, sometimes unilateral, at other times bilateral. In all cases, however, each band corresponds very closely to the distribution of the sensory fibres of a given spinal nerve. Originally supposed to be strictly a skin disease, *herpes zoster* was shown in 1861 by von Baerensprung to be an inflammation of the ganglia of the posterior nerve roots, hence its more modern name of acute posterior polyomyelitis. Since this disease is a disturbance in the sensory fibres of the spinal nerves, is it not possible that its obvious external symptom, the inflamed skin over the area of distribution of the given nerve, is due to abnormal secretions from the integumentary nerve terminals induced by the deeper nerve disturbance? Such an interpretation, though purely hypothetical, shows a possible application of the neuro-humoral principle already advanced.

Although at present there is no experimental evidence by which neuro-humoral performance in strictly receptor activities, as is implied in Botezat's hypothesis, can be demonstrated, there is no obstacle to such an application. From the standpoint of the trophic activities of receptor neurones, however, there is already much recorded that indicates, without being finally conclusive, that neuro-humoral activities pervade the receptor portion of the nervous system. This evidence distinctly favours the application of neuro-humoralism to this part of the nervous system.

Adjustors. The adjustors or central nervous organs, as already explained, are differentiated from the nerve-nets of the lower invertebrates. In these nerve-nets the primitive nerve cells or protoneurones are apparently

directly connected with one another, and the transmission of nerve impulses, if impulses may be attributed
to so simple a nervous organization, may occur in any
direction; in other words, the nerve-nets are unpolarized.
From such a system were differentiated the central
nervous organs of the higher animals. In these organs
the specialized nerve cells or neurones are not continuous as the cells in nerve-nets are, but are related to
one another through synapses, minute areas of contact

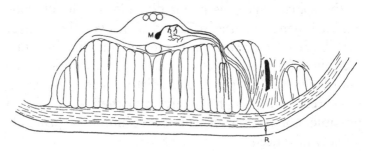

Fig. 16. Cross-section of the ventral nervous system of an earthworm
showing the relations of a receptor neurone, R, to a motor neurone, M.
From Retzius, *Biol. Unters.* N.F. **3**, taf. v, fig. 1.

that allow impulses to pass in a given direction but not
in the opposite one. Such a nervous system is polarized
and the polarizing structures are the synapses.

In the central nervous organs of so simple an animal
as an earthworm, where a receptor neurone impinges
directly upon a motor one (Fig. 16), how does the receptor half in the synapse affect the motor half that a
nerve impulse may pass from one to the other and what
is the device by which a nerve impulse is limited to one
direction only? It is now coming to be well recognized
that a nerve impulse is a wave of ionic adjustment that
passes over the fibrous portion of a neurone to discharge

at a synapse into a neighbouring neurone. How this transfer from one neurone to another is accomplished has been discussed recently by Gerard (1931, p. 74), who states that "either the same kind of ion migration and chemical response which represents successive activation of one region of the nerve fibre by another must also take place at the synapse, or it is conceivable that the end of the axon acts as a miniature gland and, when stimulated, produces some chemical which is able to excite an adjacent or neighbouring dendrite". The second alternative, which is the one that seems, in general, to be gaining in acceptance and which as early as 1925 was suggested by Sherrington as the solution of many of the difficulties in adjustor physiology, is an obvious application of the neuro-humoral principle to the conditions of the central organs. By this application it is easy to understand the polarized condition of the synapses, for when a nerve impulse arrives under normal conditions at a synaptic junction it should set up there according to the neuro-humoral interpretation a momentary discharge of secretion which on coming in contact with that segment of the synapse that belongs to the next neurone would excite in that part a new nerve impulse. Should an impulse arrive in the reverse direction, the segment of the synapse first met, being merely receptive in nature, would be incapable of secretion and hence the impulse would die out at that point without passing to the next neurone. Thus, the neuro-humoral principle offers a simple explanation not only for the polarization of the synapse but also for the retardation that the impulse is known to suffer in its passage from one neurone to another, for secretion and

excitation by secretion must be relatively time-con-
suming processes. Direct evidence of such operations is
not easily to be had, for the synapses are structurally so
minute that experimental treatment of them seems to be
out of the question.

If the neuro-humoral conception is correct and the
two sides of a synapse could be investigated, it is fair to
expect that the secretory side would appear structurally
different from the receptive side. The direct study of the
synapse has already been attempted by Bartelmez (1915)
in his investigation of the Mauthner's cells in the
teleosts. Here in consequence of the enormous size of
the cell and its processes the relations of its dendrites
to the impinging branches from other neurones, chiefly
from the eighth nerve, may be directly observed. These
synaptic junctions exhibit a plasma membrane on the
side of each neurone and in close contact one with the
other. Such are the obvious structural conditions of the
synapse, but nothing in particular has been recorded by
Bartelmez on the polarization of the junction. In certain
still larger nervous elements, the so-called giant fibres
of the invertebrates, this feature can be made out.

Giant fibres were known to many of the older investigators of the invertebrate nervous system. In 1861
Claparède noted them in the central nervous organs of
the oligochaete annelids. They were subsequently described in many other annelids and in crustaceans
(Krieger, 1880). In a transverse section of the ventral
nerve chain of an earthworm three such fibres can be
seen, one median and two lateral (Fig. 16); in the crayfish four are to be observed, two median and two lateral.
Many opinions were expressed by the older workers as

to the functions of these fibres. They were supposed by some to be nutritive tubes, by others to be blood-vessels, by still others supporting organs like the vertebrate notochord, and some even declared that they were functionless. Leydig in 1864 described them as nervous and this opinion, which was espoused by a number of other investigators, was eventually confirmed by experiment. When in the earthworm they were cut without severing the rest of the nervous cord the quick contraction of the whole worm from vigorous stimulation of the head or of the tail stops at the cut (Friedlander, 1888, 1891, 1894; Bovard, 1918; Yolton, 1923). Their function in the annelids, then, is concerned with the initiation of the quick general movements of the worm as a whole.

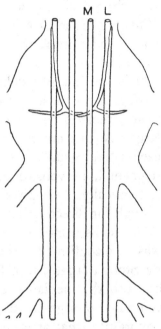

Fig. 17. Diagram of the giant fibres in the crayfish *Cambarus* seen in their passage through the fifth abdominal ganglion: *M*, median fibres, continuous; *L*, lateral fibres, segmented and showing an overlap, macrosynapse, in the anterior portion of the ganglion. From Johnson, *Jour. Comp. Neurol.* 1924, **36**, 355.

Before 1924 all investigators were united in the opinion that each giant fibre was a single continuous structure extending throughout the greater part of the length of the animal's nerve cord. In that year, however, Johnson showed that in certain crustaceans, *Cambarus* and *Palaemonetes*, though the median fibres were continuous single structures, the lateral ones were composed of a series of overlapping elements each of which was one segment long and so applied to the elements in front and behind that the whole appeared as a single continuous body (Fig. 17). It is probable that each segment in such a compound fibre represents a single neurone and since the fibre as a whole conducts, the areas of overlap, which are sharply delimited by plasma membranes analogous to those already described by Bartelmez in fishes, must be a synapse. Because of its unusually large size this might be called a macrosynapse. Thus, each lateral giant fibre in a crustacean is a series of segmental, nervous units, probably neurones, united anatomically and physiologically by synapses large enough for direct observation.

Johnson's work on the giant fibres in the crustaceans was soon confirmed in the earthworm and other annelids by Stough (1926), who found that not only the lateral but also the median giant fibres of these forms were composed of segmented units and, what was of greater importance, that at each overlap or macrosynapse the protoplasmic substance of the element of one side always stained in osmic acid distinctly deeper than that of the other side (Fig. 18). Thus, a difference in the staining capacity of the elements on the two sides of the synapse was for the first time demonstrated and in this way gave visible expression to synaptic polarization. The difference in the staining capacities of the two sides of the macro-synapse which was thus demonstrated was sought for in the lateral giant fibres of crustaceans by Warner (1931), who identified it in appropriate positions in the nerve cord of the crayfish. Thus, in both annelids and crustaceans the segmented giant fibres exhibit histological polarization in that the opposed sides of each macro-synapse respond differently to such reagents as osmic acid.

Since the publication of these discoveries both Johnson (1926) and Stough (1930) have worked experimentally on the direction of transmission in the giant fibres.

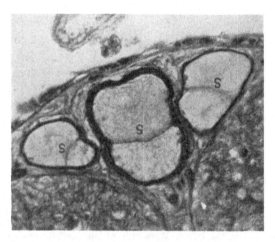

Fig. 18. Cross-section of the three giant fibres in the ventral nerve cord of the earthworm, showing in each instance both members of an overlap. In all cases the more deeply stained member of the pair is above. The membrane S, probably double, that separates the two members represents the macrosynapse. Stough, *Jour. Comp. Neurol.* 1926, **40**, 449.

Johnson showed that in crustaceans the median fibres transmit antero-posteriorly and the lateral ones in the reverse direction. Stough's results on the earthworm agree with those of Johnson. When in an earthworm the median giant fibre is cut antero-posterior transmission is interrupted at the break but postero-anterior transmission is undisturbed. When the lateral fibres are cut antero-posterior transmission is undisturbed while postero-anterior is blocked at the cut. Thus, in both animals the median fibres transmit toward the tail, the lateral toward the head. The giant fibres then show not only histological but also physiological polarization.

If now these types of polarization are set forth in one scheme, as shown in Fig. 19, it will be observed that at

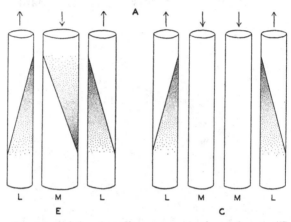

Fig. 19. Diagrams of the giant fibre systems in the earthworm (*E*) and in the crayfish (*C*). Their macrosynapses, in all three fibres of the earthworm and in the lateral fibres only of the crayfish, are indicated by oblique partitions and the deeply staining member of each synapse is appropriately shaded. *A*, anterior; *L*, lateral fibre; *M*, median fibre. Arrows indicate the direction of transmission of nervous impulses. It will be seen that in each macrosynapse the deeply stained member is the discharging one and the faintly stained member the receptive one. The diagram is based on the work of Johnson, 1924, 1926, Stough, 1926, 1930, and Warner, 1931.

every macrosynapse the deeply stained element is on the discharging side of the combination and the faintly stained one on the receptive side. Interpreted in terms of neuro-humoral activity the deeply stained side of the macrosynapse is the secretory side and the lightly stained one is the receptive side. The evidence thus put forward demonstrates beyond a doubt a chemical difference in the two sides of the macrosynapse, but it is, of course, not a demonstration of secretory activity. What it does show, however, is that the neuro-humoral interpretation may be applied without difficulty to the synaptic nervous system, and that by this means, as Sherrington and others have suggested, many adjustor problems may be successfully met.

Conclusion. The extent to which the neuro-humoral hypothesis applies to the nervous system in general is a problem for future investigation. That it is the underlying principle in the relation of the excitatory nerves and the simpler types of chromatophores is beyond dispute. Probably this condition holds also for glands and muscles, but the extent to which such a view is true for these and other effectors is still to be ascertained. In receptor and adjustor activities important instances occur such as the trophic action of receptor neurones and the polarization of giant fibres, which show how readily this principle can be made to apply to these aspects of nervous organization. But that neuro-humoralism is the solution of the various problems involved in these fields is still to be demonstrated. Much of this uncertainty is due to the novelty of this view, a situation that can be changed only by extended investigation. What the outcome of such work may be is impossible to predict, but that the little that is known is favourable to neuro-humoralism is not without interest. In many respects this principle is a special application of Bayliss and Starling's concept of hormones, though the distances travelled by the substances from their regions of origin to those of application are often extremely short. In another way it is a detailed illustration of Claude

Bernard's conception of the "milieu intérieur" and it harks back to a still older school of French physiologists headed by the illustrious Cabanis. Nobody to-day considers seriously the aphorism that "the brain secretes thought as the liver secretes bile", but if what has been set down in the preceding pages be true, nervous secretion plays a part even in our mental operations such as has scarcely been suspected by the modern physiologist.

REFERENCES

ADLER, L. (1914). Metamorphosestudien an Batrachierlarven. *Arch. Entw.-mech. Organ.* **39**, 21–45.

ALLEN, B. M. (1917). Effects of the extirpation of the anterior lobe of the hypophysis of *Rana pipiens*. *Biol. Bull.* **32**, 117–130.

—— (1930). Sources of the pigmentary hormone of amphibian hypophysis. *Proc. Soc. Exper. Biol. Med.* **27**, 504–505.

ATWELL, W. J. (1919). On the nature of the pigmentation changes following hypophysectomy in the frog larva. *Science*, **49**, 48–50.

VON BAERENSPRUNG, F. G. F. (1861). Die Gürtelkrankheit. *Ann. Char.-Krankenh. Berlin*, **9**, 40–128.

BALLOWITZ, E. (1893). Die Nervenendigungen der Pigmentzellen, ein Beitrag zur Kenntnis des Zusammenhanges der Endverzweigungen der Nerven mit dem Protoplasma der Zellen. *Zeit. wiss. Zool.* **56**, 673–706.

BARTELMEZ, G. W. (1915). Mauthner's cell and the nucleus motorius tegmenti. *Jour. Comp. Neurol.* **25**, 87–128.

BAUER, V. (1905). Ueber einen objectiven Nachweis des Simultancontrastes bei Tieren. *Centralbl. Physiol.* **19**, 453–462.

BAUER, V., und E. DEGNER (1913). Ueber die allgemein-physiologische Grundlage des Farbenwechsels bei dekapoden Krebsen. *Zeit. allg. Physiol.* **15**, 363–412.

BAYLISS, W. M., and E. H. STARLING (1902). The mechanism of pancreatic secretion. *Jour. Physiol.* **28**, 325–353.

BERT, P. (1875). Sur le mécanisme et les causes des changements de couleur chez le Caméléon. *C.R. Acad. Sci. Paris*, **81**, 938–941.

BIEDERMANN, W. (1892). Ueber den Farbenwechsel der Frösche. *Arch. ges. Physiol.* **51**, 455–508.

BOTEZAT, E. (1909). Die sensiblen Nervenendapparate in den Hornpapillen der Vögel im Zusammenhang mit Studien zur vergleichenden Morphologie der Sinnesorgane. *Anat. Anz.* **34**, 449–468.

—— (1910). Ueber Sinnesdrüsenzellen und die Funktion von Sinnesapparaten. *Anat. Anz.* **37**, 513–530.

BOVARD, J. F. (1918). The function of the giant fibers in earthworms. *Univ. California Pub., Zool.* **18**, 135–144.

BRINKMAN, R., und E. VAN DAM (1922). Die chemische Uebertragbarkeit der Nervenreizwirkung. *Arch. ges. Physiol.* **196**, 66–82.

BRINKMAN, R., und M. RUITER (1924). Die humorale Uebertragung der neurogenen Skelettmuskelerregung auf den Darm. *Arch. ges. Physiol.* **204**, 766–768.

BRINKMAN, R., und M. RUITER (1925). Die humorale Uebertragung der Skelettmuskelreizung eines ersten auf den Darm eines zweiten Frosches. *Arch. ges. Physiol.* **208**, 58–62.

BROCKELBANK, M. C. (1925). Degeneration and regeneration of the lateral-line organs in *Ameiurus nebulosus* (Les.). *Jour. Exp. Zool.* **42**, 293–305.

BRÜCKE, E. (1852). Untersuchungen ueber den Farbenwechsel des afrikanischen Chamäleons. *Denkschr. Akad. Wiss. Wien, math.- nat. Cl.* **4**, 179–210.

CANNON, W. B., and Z. M. BACQ (1931). Studies on the conditions of activity in endocrine organs. XXVI. A hormone produced by sympathetic action on smooth muscle. *Amer. Jour. Physiol.* **96**, 392–412.

CHARLES, E. (1931). Metabolic changes associated with pigmentary effector activity and pituitary removal in *Xenopus laevis*. II. Calcium and magnesium content of the serum. *Proc. Roy. Soc. London.* B, **107**, 504–510.

CLAPARÈDE, E. (1861). Études anatomiques sur les annélides, turbellariés, opalines et grégarines observés dans les Hébrides. *Mém. Soc. Phys. Hist. Nat. Genève*, **16**, 73–164.

CORONA, A., e A. MORONI (1898). Contributo allo studio dell' estratto di capsuli surrenali. *La Reforma Medica*, **14** (cited from van Rynberk, 1906).

DALE, H. H., and J. H. GADDUM (1930). Reactions of denervated voluntary muscle, and their bearing on the mode of action of parasympathetic and related nerves. *Jour. Physiol.* **70**, 109–144.

DEGNER, E. (1912a). Ueber Bau und Funktion der Kruster-chromatophoren. *Zeit. wiss. Zool.* **102**, 1–78.

—— (1912b). Weitere Beiträge zur Kenntnis der Crustaceenchromatophoren. *Zeit. wiss. Zool.* **102**, 701–710.

DEMOOR, J. (1913). Le mécanisme intime de la sécrétion salivaire. *Arch. intern. Physiol.* **13**, 187–206.

—— (1929). Le réglage humoral dans le cœur. *Ann. Physiol. Physiochémie Biologique*, **5** (no. 1).

ELLIOTT, T. R. (1904). On the action of adrenalin. *Jour. Physiol.* **31**, xx–xxi.

—— (1905). The action of adrenalin. *Jour. Physiol.* **32**, 401–467.

FLOROVSKY, G. B. (1917). On the mechanism of reflex salivary secretion. *Bull. Acad. Imp. Sci. Petrograd*, 1917, 119–136.

FORTUYN, A. B. D. (1920). *Die Leitungsbahnen im Nervensystem der wirbellosen Tiere.* Haarlem, 370 pp.

FREDERICQ, H. (1925). Le mécanisme humoral des actions nerveuses. *Rev. Sci.* **63**, 641–648.

—— (1927). La transmission humorale des excitations nerveuses. *C.R. Soc. Biol. Paris*, **96** (supplement), 3–38.

FRIEDLANDER, B. (1888). Beiträge zur Kenntnis des Centralnervensystems von Lumbricus. *Zeit. wiss. Zool.* **47**, 47–84.

FRIEDLANDER, B. (1891). Ueber die markhaltigen Nervenfasern und Neurochorde der Crustaceen und Anneliden. *Mitth. zool. Stat. Neapel*, **9**, 205–265.

—— (1894). Beiträge zur Physiologie des Centralnervensystems und des Bewegungsmechanismus der Regenwürmer. *Arch. ges. Physiol.* **58**, 168–206.

FRIES, E. F. B. (1927). Nervous control of xanthophore changes in *Fundulus*. *Proc. Nat. Acad. Sci.* **13**, 567–569.

VON FRISCH, K. (1910). Ueber die Beziehungen der Pigmentzellen in der Fischhaut zum sympathischen Nervensystem. *Festschrift R. Hertwig*, **3**, 15–28.

—— (1911a). Beiträge zur Physiologie der Pigmentzellen in der Fischhaut. *Arch. ges. Physiol.* **138**, 319–387.

—— (1911b). Die Pigmentzellen der Fischhaut. *Biol. Centralbl.* **31**, 412–415.

—— (1912). Ueber farbige Anpassung bei Fischen. *Zool. Jahrb., Abt. allg. Zool. Physiol.* **32**, 171–230.

FRÖHLICH, A. (1910). Farbwechselreaktion bei *Palaemon*. *Arch. Entw.-mech. Organ.* **29**, 432–438.

GERARD, R. W. (1931). Nerve conduction in relation to nerve structure. *Quart. Rev. Biol.* **6**, 59–83.

GIERSBERG, H. (1930a). Der Farbwechsel der Fische. *Zeit. vergl. Physiol.* **13**, 258–279.

—— (1930b). Der Farbenwechsel der Tiere. *Jahrb. Schles. Gesell. vaterl. Cultur*, **103**, 1–11.

—— (1930c). Der Farbwechsel der Tiere. *Forschungen und Fortschritte*, **6**, 450–451.

GRIFFINI, L. (1887). Sulla riproduzione degli organi gustatorii. *Rend. real. Ist. Lombardo Sci. Lett.* ser. 2, **20**, 667–683.

HANSTRÖM, B. (1928). *Vergleichende Anatomie des Nervensystems der wirbellosen Tiere, unter Berücksichtigung seiner Funktion.* Berlin, 628 pp.

HEWER, H. R. (1926). Studies in colour changes of fish. II. An analysis of the colour patterns of the dab. *Phil. Trans. Roy. Soc. London*, B, **215**, 177–186.

HOGBEN, L. T. (1924). *The pigmentary effector system.* Edinburgh and London, 152 pp.

HOGBEN, L. T., and L. MIRVISH (1928a). Some observations on the production of excitement pallor in reptiles. *Trans. Roy. Soc. S. Africa*, **16**, 45–52.

—— —— (1928b). The pigmentary effector system. V. The nervous control of excitement pallor in reptiles. *British Jour. Exp. Biol.* **5**, 295–308.

HOGBEN, L. T., and D. SLOME (1931). The pigmentary effector system. VI. The dual character of endocrine co-ordination in amphibian colour change. *Proc. Roy. Soc. London*, B, **108**, 10–53.

HOWELL, W. H., and W. W. DUKE (1908). The effect of vagus inhibition on the output of potassium from the heart. *Amer. Jour. Physiol.* **21**, 51–63.

JOHNSON, G. E. (1924). Giant nerve fibers in crustaceans with special reference to *Cambarus* and *Palaemonetes*. *Jour. Comp. Neurol.* **36**, 323–373.

—— (1926). Studies on the functions of the giant fibers of crustaceans, with special reference to *Cambarus* and *Palaemonetes*. *Jour. Comp. Neurol.* **42**, 19–33.

KAPPERS, C. U. A. (1929). *The evolution of the nervous system in invertebrates, vertebrates and man.* Haarlem, 335 pp.

KEEBLE, F. W., and F. W. GAMBLE (1900). The colour-physiology of *Hippolyte varians*. *Proc. Roy. Soc. London*, **65**, 461–468.

KOLLER, G. (1925). Farbwechsel bei *Crangon vulgaris*. *Verh. deutsch. zool. Ges.* **30**, 128–132.

—— (1927). Ueber Chromatophorensystem, Farbensinn und Farbwechsel bei *Crangon vulgaris*. *Zeit. vergl. Physiol.* **5**, 191–246.

—— (1928). Versuche ueber die inkretorischen Vorgänge beim Garneelenfarbwechsel. *Zeit. vergl. Physiol.* **8**, 601–612.

—— (1929). Die innere Sekretion bei wirbellosen Tieren. *Biol. Rev.* **4**, 269–306.

—— (1930). Weitere Untersuchungen ueber Farbwechsel und Farbwechselhormone bei *Crangon vulgaris*. *Zeit. vergl. Physiol.* **12**, 632–667.

KOLLER, G., und E. MEYER (1930). Versuche ueber den Wirkungsbereich von Farbwechselhormonen. *Biol. Zentralbl.* **50**, 759–768.

KRIEGER, K. R. (1880). Ueber das Centralnervensystem des Flusskrebses. *Zeit. wiss. Zool.* **33**, 527–594.

KROPP, B. (1927). The control of the melanophores in the frog. *Jour. Exp. Zool.* **49**, 289–318.

—— (1929). The melanophore activator of the eye. *Proc. Nat. Acad. Sci. Washington*, **15**, 693–694.

KRÖYER, H. (1842). Monographisk Fremstilling af Slaegten Hippolyte's Nordiske Arten. *Kong. Dansk. Videnskap. Selskabet*, **9**, 209–361.

KUNTZ, A. (1917). The histological basis of adaptive shades and colours in the flounder *Paralichthys albiguttus*. *Bull. United States Bur. Fish.* **35**, 5–29.

LEHMANN, C. (1923). Farbwechsel bei *Hyperia galba*. *Biol. Zentralbl.* **43**, 173–175.

LEYDIG, F. (1864). *Vom Bau des thierischen Körpers.* Tübingen, 278 pp.

LIEBEN, S. (1906). Ueber die Wirkung von Extrakten chromaffinen Gewebes (Adrenalin) auf die Pigmentzellen. *Centralbl. Physiol.* **20**, 108–117.

LOEWI, O. (1921). Ueber humorale Uebertragbarkeit der Herznervenwirkung. *Arch. ges. Physiol.* **189**, 239–242.

MAST, S. O. (1916). Changes in shade, colour, and pattern in fishes, and their bearing on the problems of adaptation and behavior, with especial reference to the flounders *Paralichthys* and *Ancylopsetta*. *Bull. United States Bur. Fish.* 34, 173–238.

MAY, R. M. (1924). Skin grafts in the lizard, *Anolis Carolinensis* Cuv. *British Jour. Exp. Biol.* 1, 539–559.

—— (1925). The relation of nerves to degenerating and regenerating taste-buds. *Jour. Exp. Zool.* 42, 371–410.

MAYER, P. (1879). Ueber Farbenwechsel bei Isopoden. *Mitth. zool. Stat. Neapel*, 1, 521–522.

MEGUŠAR, F. (1912). Experimente ueber den Farbwechsel der Crustaceen. *Arch. Entw.-mech. Organ.* 33, 462–665.

MENKE, H. (1911). Periodische Bewegungen und ihr Zusammenhang mit Licht und Stoffwechsel. *Arch. ges. Physiol.* 140, 37–91.

MEYER, E. (1930). Ueber die Mitwirkung von Hormonen beim Farbwechsel der Fische. *Forschungen und Fortschritte*, 6, 379–380.

MEYER, S. (1896). Durchschneidungsversuche am Nervus glossopharyngeus. *Arch. mik. Anat.* 48, 143–145.

OLMSTED, J. M. D. (1920). The nerve as a formative influence in the development of taste-buds. *Jour. Comp. Neurol.* 31, 465–468.

PARKER, G. H. (1909). The origin of the nervous system and its appropriation of effectors. *Pop. Sci. Monthly*, 75, 56–64, 137–146, 253–263, 338–345.

—— (1910). The reactions of sponges, with a consideration of the origin of the nervous system. *Jour. Exp. Zool.* 8, 1–41.

—— (1919). *The elementary nervous system*. Philadelphia and London, 229 pp.

—— (1923). The origin and development of the nervous system. *Scientia*, 34, 23–32.

—— (1930). The colour changes of the tree toad in relation to nervous and humoral control. *Proc. Nat. Acad. Sci.* 16, 395–396.

PERKINS, E. B. (1928). Colour changes in crustaceans, especially in *Palaemonetes*. *Jour. Exp. Zool.* 50, 71–105.

PIÉRON, H. (1913). Le mécanisme de l'adaptation chromatique et la livrée nocturne de l'*Idotea tricuspidata* Desm. *C.R. Acad. Sci. Paris*, 157, 951–953.

—— (1914). Recherches sur le comportement chromatique des invertébrés et en particulier des Isopodes. *Bull. Sci. France et Belg.* 48, 30–79.

POUCHET, G. (1872a). Du rôle des nerfs dans les changements de coloration des poissons. *Jour. Anat. Physiol.* 8, 71–74.

—— (1872b). Sur les rapides changements de coloration provoqués expérimentalement chez les crustacés et sur les colorations bleues des poissons. *Jour. Anat. Physiol.* 8, 401–407.

—— (1876). Des changements de coloration sous l'influence des nerfs. *Jour. Anat. Physiol.* 12, 1–90, 113–165.

RANVIER, L. (1882). *Traité technique d'histologie*. Sixième fasc. Paris.

REDFIELD, A. C. (1918). The physiology of the melanophores of the horned toad *Phrynosoma*. *Jour. Exp. Zool.* **26**, 275–333.

RETZIUS, G. (1890). Zur Kenntnis des Nervensystems der Crustaceen. *Biol. Unters.* N.F. **1**, 1–50.

SANDMEYER, W. (1895). Ueber das Verhalten der Geschmacksknospen nach Durchschneidung des Nervus glossopharyngeus. *Arch. Anat. Physiol., physiol. Abt.*, 1895, 269–276.

SARS, G. (1867). *Histoire naturelle des Crustacés d'eau douce de Norvège*. Christiania, 145 pp.

SCHLIEPER, C. (1926). Der Farbwechsel von *Hyperia galba*. *Zeit. vergl. Physiol.* **3**, 547–557.

SHARPEY-SCHAFER, E. (1924). *The Endocrine Organs*, Part I. London, 175 pp.

SHERRINGTON, C. S. (1925). Remarks on some aspects of reflex inhibition. *Proc. Roy. Soc. London*, B, **97**, 519–545.

SLOME, D., and L. HOGBEN (1928). The chromatic function in *Xenopus laevis*. *South African Jour. Sci.* **25**, 329–335.

SMITH, D. C. (1930). The effect of temperature changes upon the chromatophores of crustaceans. *Biol. Bull.* **58**, 193–202.

—— (1931). The action of certain autonomic drugs upon the pigmentary responses of *Fundulus*. *Jour. Exp. Zool.* **58**, 423–453.

SMITH, P. E. (1916). The effect of hypophysectomy in the early embryo upon the growth and development of the frog. *Anat. Rec.* **11**, 57–64.

SPAETH, R. A. (1916). Evidence proving the melanophore to be a disguised type of smooth muscle cell. *Jour. Exp. Zool.* **20**, 193–215.

SPAETH, R. A., and H. G. BARBOUR (1917). The action of epinephrin and ergotoxin upon single, physiologically isolated cells. *Jour. Pharm. Exp. Therap.* **9**, 431–440.

SPEIDEL, C. C. (1919). Gland-cells of internal secretion in the spinal cord of the skates. *Publ. Carnegie Inst. Washington*, **281**, 1–31.

—— (1922). Further comparative studies in other fishes of cells that are homologous to the large irregular glandular cells in the spinal cord of the skates. *Jour. Comp. Neurol.* **34**, 303–317.

STARK, J. (1830). On changes observed in the colour of fishes. *Edinburgh New Philos. Jour.* **23**, 327–331.

STOUGH, H. B. (1926). Giant nerve fibers in the earthworm. *Jour. Comp. Neurol.* **40**, 409–463.

—— (1930). Polarization of the giant nerve fibers of the earthworm. *Jour. Comp. Neurol.* **50**, 217–231.

SUMNER, F. B. (1911). The adjustment of flat-fishes to various backgrounds. *Jour. Exp. Zool.* **10**, 409–505.

SWINGLE, W. W. (1921). The relation of the pars intermedia of the hypophysis to the pigmentation changes in anuran larvae. *Jour. Exp. Zool.* **34**, 119–141.

TAITE, J. (1910). Colour changes in the isopod *Ligia oceanica*. *Jour. Physiol.* **40**, xl–xli.

TEN CATE, J. (1924). Sur la question de l'action humorale du nerf vague. *Arch. néerl. Physiol.* **9**, 588–597.

VALLISNIERI, A. (1715). *Osservazioni intorno alle rane Venezia* (cited from van Rynberk, 1906).

VON VINTSCHGAU, M., and J. HÖNIGSCHMIED (1876). Nervus glossopharyngeus und Schmeckbecher. *Arch. ges. Physiol.* **14**, 443–448.

WARNER, S. G. (1931). Histological polarization of lateral giant fibers in the crayfish. *Proc. Nat. Acad. Sci.* **17**, 140–141.

WYMAN, L. C. (1922). The effect of ether upon the migration of the scale pigment and the retinal pigment in the fish, *Fundulus heteroclitus*. *Proc. Nat. Acad. Sci.* **8**, 128–130.

—— (1924). Blood and nerve as controlling agents in the movements of melanophores. *Jour. Exp. Zool.* **39**, 73–132.

YOLTON, L. W. (1923). The effects of cutting the giant fibers in the earthworm, *Eisenia foetida* (Sav.). *Proc. Nat. Acad. Sci.* **9**, 383–385.

Printed in the United States
By Bookmasters